Mules to Rockets

Mules
to
Rockets

A Memoir

Charles H. Jackson

Tate Publishing & *Enterprises*

Published by Tate Publishing & Enterprises, LLC
127 E. Trade Center Terrace | Mustang, Oklahoma 73064 USA
1.888.361.9473 | www.tatepublishing.com

Tate Publishing is committed to excellence in the publishing industry. The company reflects the philosophy established by the founders, based on Psalm 68:11,
"The Lord gave the word and great was the company of those who published it."

Cover design by April Marciszewski
Interior design by Joel Uber

Published in the United States of America

ISBN: 978-1-61777-732-5
Biography & Autobiography: Personal Memoirs
11.06.29

Dedication

This book is dedicated to my parents, Fletcher and Mattie Jackson; my wife, Judi; my children, Jana, Kevin, Alicia; and my grandchildren, Emily, John, Hannah, and Braden.

Acknowledgments

A special thanks to my wife, Judi, and my granddaughter, Emily, for some of the artwork, editing, and for their patience with me.

Table of Contents

Preface

This is an autobiography of events of my life, documented primarily for my children, grandchildren, and their descendants, to enable them to understand the period of time in which I lived. It also includes some insight into some national historic events, not heretofore revealed, in which I was involved.

I have intentionally selected events and happenings in my life that would make the reading interesting. It was not intended that I tell every little detail of my life for fear of boring the reader to the point of tears.

Some of the material of this book may appear that I am bragging, but I assure you that is not the intent. The purpose is to record for posterity.

The people brought into the stories documented herein are real, and it is my wish and hope that none will be offended by the use of their name, because my intentions were only good.

Philosophy

There in my recliner, in the fully reclined position and somewhere between consciousness and sleep, I was mulling over the events of my life. I began to realize that I had seen and been involved in a lot of history, spanning from the time where the primary mode of transportation for a rural family was a mule or horse-drawn wagon, to rocket travel by astronauts to the moon and a space station.

It became apparent to me that my life was quite varied and, in some cases, unique and that I was really the only one that knew the whole story. I realized that no one else would ever know the story unless it was recorded for them in some way. Having three children and four grandchildren that might someday be interested in their ancestry, I decided to record some events of my life. After thinking about those events for a while, I thought that it would be better to not tell everything, lest I start out by going to a church and by saying again, "Father, forgive me again."

I have noted that when people try to recall events from the early years of their life, even with prompting from adults that witnessed those events, they hardly ever have full recall before ages four or five. Thinking back, I hardly remember anything before the age of four. The things I do remember before that age are merely snippets of events, and only then those that made a monumental impression on me at the time.

I took a psychology course in college and another during my working career where the professors talked of the various phases of human life and how, in the early years, we are basically only an extension of our parents, soaking up information at an astounding rate. When that information reaches a certain level of concentration in our brains, then we become aware that we are an individual, separate from our parents, and capable of thinking somewhat on our own. In those courses I also learned that the human brain continues to develop, so far as making good decisions, until the early twenties. I suppose that is why we often see more teenage accidents than with adults. Obviously, many never get out of their early twenties mentally.

In my estimation, guidance from adults is necessary until a child's brain is fully developed and capable of making good, responsible decisions. I personally believe that many of our societal problems today are caused by the lack of adult, primarily parental, guidance. That is why we see so many problem children from broken homes.

I believe that everyone's life on this earth is an adventure beginning at birth and ending when our earthly body dies. I believe that we continue learning our whole life, building on, and making relationships with the previous knowledge we store in our brains. Whether the adventure continues on with our creator after the earthly death or not is up to each of us to choose.

Beginning

Okay, here we go. So just hold on. I was born July 10, 1936, just south of Tuscumbia, Alabama, as the only son of a tenant farmer. Tuscumbia is known as the birth place of Helen Keller, the home of the late US Senator Howell Heflin, and the western terminus of the first railroad, started in 1830, on the American frontier. Tuscumbia is built near a big spring in the beautiful foothills of the Appalachian Mountains.

The French first settled the area near the Tennessee River in the 1700s only to have their town destroyed in 1786. The Chickasaw Indians moved there next, building their own town near the big spring. They were soon followed by the Michael Dickson family around 1815.

The Federal Government built a road from Nashville, Tennessee, to Tuscumbia in 1817 to 1819, opening up the town, which eventually led to its incorporation as Ococoposa (cold water) in 1820. The town later changed its name to Big Spring and in 1822, a town vote gave it the name Tuscumbia, after a Chickasaw Indian chief, who was known for giving a pair of moccasins to the first white child born in the area.

My parents were Fletcher B. and Mattie E. (Taylor) Jackson. My ancestors were European (Scot, Irish, English, and Dutch) and Native American. My grandfather Jackson spoke of his great-grandfather with two brothers coming to America in the mid-1700s to seek their fortune and to escape the political, social, and religious oppression in Europe, as well as frequent crop famine and disease. Two of the brothers continued on west and the other brother, of my lineage, stayed in South Carolina.

My great-great grandfather, Tyre Jackson, was in north Georgia in the 1820s and '30s, during the years of the Georgia land lotteries and the gold rush. There he married an Indian maiden whose story I have recorded in my first book, *Su Ki and the Settler*.

Even though I do not remember anything of the earliest years of my life, I learned later in life that I was born at home and delivered by Dr. Montgomery, who lived a few miles away, and also delivered all three of my sisters, Doris, Dorothy, and Martha. Being born during the Great Depression, money was very scarce and doctors often accepted farm produce as payment. Dr. Montgomery was more than glad to accepted partial payment in produce.

The Great Depression was hard enough on most people at that time, but that was not the only source of misery for many tenant farmers during those days. Landlords throughout the country often held the tenants in virtual slavery by controlling almost every aspect of their lives. Just being a tenant farmer usually meant that the landlord provided the land, seed, fertilizer, mules or horses, and implements, and the tenant did all the work. A tenant usually started off indebted to the landowner because he had to live a year in advance of producing the first crops. Even after that period, the landlord usually got at least one third, and some times more, of what the tenant produced. Because of little automation, the tenant barely raised enough to live on after the landowner got his share.

Often tenants would seek other work in their spare time to supplement their income. My father was no exception, because he worked part-time, during summer months, measuring cotton acreage for the US government-sponsored allotment program. For that work he was paid a modest fifty cents per day. That doesn't sound like much money, but during the Great Depression, fifty cents would buy a twenty-five pound bag of flour, a stand of lard, or several bags of groceries.

My father was also more fortunate than most tenant farmers because his landlord was his brother-in-law, Authur Blackburn, who did not demand as much as most landlords did at that time. Even with that advantage, we were still poor. As was often said, "We were so poor that we could only afford to spell poor with one *o* and without the *r*." I also learned later in my life that my father's sister, Bertha, provided a lot of food, medicine, and hand-me-down clothes to my family, no strings attached.

Bertha and Authur had three boys who were often at our house helping my father with the farming chores. After mother gave birth to twin girls, Doris and Dorothy, in 1934, Bertha helped her find a black lady, Mary Lizzie Johnson, to help with the house work, garden, and canning of vegetables. Even then Bertha spent a lot of her time helping mother when the twins were first born. I came along in 1936 and our younger sister Martha was born in 1939. Authur helped my father hire other part-time help. Mary Lizzie's husband, Roy, and his friend Humphrey (Ump) worked for my dad during the farming season.

Age Four

I do remember a number of things beginning around age four. The house in which I was born and lived in until age six, made an impression on me. It was clapboard tenant house with four rooms. The living room and kitchen were located on one side of a "dog trot" and the other two unheated bedrooms were on the other side of the dog trot. A dog trot is an open hallway extending from the front porch of the house to the back porch; it was called this since it wasn't uncommon to find your dog and several of the neighbors' dogs lying in that open hallway to stay out of the summer heat. Many houses were built this way in the South to allow a cool breeze to blow through in the sweltering heat of summer months, since there was no air conditioning at that time. The dog trot was a place of much activity during the summer; some folks even slept there when it was extremely hot.

There was not any electricity in most rural locations until after the National Rural Electrification Program of the early 1940s, so few even had a fan.

The house also had a large front and back porch and a cellar, which was under the bedroom side of the house and could be entered by a butterfly door on the outside. There was nothing under the living room and kitchen, but chickens could get under there, because there was no underpinning to keep them out.

What I remember most is how cold the house was in the winter, when I had to leave the fireplace and cross the dog trot to get to the unheated bedroom. If it had not been for a feather mattress, lots of quilts, and some flat irons and football-sized smooth rocks that mother preheated by the fireplace, wrapped in a towel, and put in the foot of the bed, I would probably have frozen to death. Let me tell you, when you get up in the morning, get dressed, and pick up the chamber pot to take out to empty and it is frozen solid, you are not having much fun! Oh! You don't know what a chamber pot is? It's just a big pot with a lid that you use during the night when you don't have indoor restroom plumbing.

Tenant House

The cracks in the board flooring were so wide that snakes could come up into the house, which actually happened during some of the cool fall nights. More than once a snake was found lying on the warm hearth, seeking some heat from the warm stones after the fire had died down during the night. The floor cracks also provided an opportunity for some fun, because we could lie on the floor and drop corn or bread crumbs to the chickens underneath.

One event that is embedded in my mind is when I was around age four and I was chewing some honeycomb to extract the honey. My tongue was stuck by a bee stinger with poison sack attached. My tongue swelled so much that I couldn't talk or close my mouth until the next afternoon.

Several men from surrounding farms gathered at our house on a Saturday morning, with the intent to rob the honey from a beehive in a large sweet gum tree, at one end of a pond. I was not allowed to go because mother was afraid that I would get in the way or get stung by the bees. The tree was within sight of our house, and I could stand near the pasture fence and watch the activity in the creek bottom below. The men first built a scaffold around the tree, because the hive was about ten feet off the ground. Then they stood on the scaffolding and, using a two man cross cut saw, cut the tree about a foot above the bee hole.

There was a lot of hollering when the bees got excited and started swarming around the men. Being ill-protected against the onslaught of the bees, the men got many stings and one even fell off the scaffolding into the muddy bog. After four hours of hard work and many bee stings, the men walked back to the house with a tub full of honey. They divided the honey and departed for their homes to treat their stings.

Hog killing is another event that stuck in my mind. Watching preparation for the killing and butchering of hogs was a major happening for me. My father, along with hired help, started out by heating large pots of water to boiling. They then killed the hog and placed it in a wooden tank lined with metal. Next, they poured the boiling water over the hog to enable the hair to be easily removed. The hog was then hung from a scaffold by its back legs, gutted, and scraped of its hair using large knives. It was then laid out on a large table made of planks and butchered into major cuts such as, hams, middling, etc., and then put into salt boxes for curing. Minor pieces and scraps were ground and seasoned for sausage.

After curing in the salt boxes for many days, the meat was removed, washed to remove most of the salt, and then hung in a smokehouse, along with the sausage, for further curing over a smoldering hickory wood fire.

I doubt if there are many people alive today that can remember walking through a dog trot and getting a whiff of meat curing in salt boxes and from large crocks full of aging kraut. It is a very distinctive smell that you never forget.

At this point, my older twin sisters had started to school. As most people know, every disease that gets to school is carried home by most of the students. Such was the case at my home, for I got German measles and mumps, both within a span of a few months. When I had the mumps, Mother told me to not stand by the cold window or the mumps might fall on me. She said mumps could fall on boys, you can guess where, and could cause severe swelling. I remember checking myself almost daily to see if those mumps were dropping.

My parents had three African-Americans working for them: Roy and Mary Lizzie were husband and wife, and Humphrey (Ump) was a single man. One day, my three sisters and I were playing on a full cotton wagon when Mother told us to get down. Ump was standing nearby and offered to help us down. When we got in the house, mother noticed that Martha had some black grease on her dress from the wagon wheel axle. When asked where the stain came from, Martha replied without hesitation, "Ump got it on me!" not realizing that his dark color would not rub off. Mother and the rest of us laughed loud enough that Ump heard it all. Ump was a good-natured man and took no offense to Martha's lack of understanding.

Around our house, segregation was an unknown. That is not to say that segregation did not exist, for it did. As children, we did know that the blacks usually did not drink from the same dipper in the water bucket, but we didn't know why and never thought to ask. However, when we ate meals, Mary Lizzie would sit at the table with us. It was not until I was in school that I became aware of segregation, what it was, and why it existed. In the South, we were taught that some landowners after the Civil War had land taken from them by federal agents, whether they had slaves or not, and given to the freed slaves. Two generations removed from the Civil War there was still a lot of hatred and animosity toward African-Americans and northerners for that reason. That is why a lot of southern folks referred to northerners as "damn Yankees." That scenario did not make segregation right, but that was just the way it was at that time.

Utilities in the tenant house were nonexistent. There was a cistern by the back porch with a hand pump where a water bucket could be filled. The water bucket was kept in the kitchen or on the back porch and had a common dipper for use.

The stove was a wood burner and was used for all the cooking, baking, and heating water. Providing wood for the stove was a year-round chore. The only heat in the living room was a wood-burning fireplace, which required a lot of work to maintain. There was firewood to cut in the summer and fall so that it would dry before winter, and bringing the wood into the house and taking the ashes out in the winter was a daily chore.

There were no bathroom facilities at all. We had baths about once a week in the kitchen using a number two or three galvanized steel tub that was also used to wash clothes. Water was heated on the stove in a kettle. The rest of the time, we used a pan of water and a washcloth for bathing. The soap we used was called "lye soap" and was usually homemade using lye extracted from wood ashes to convert animal fat in much the same way soap is commercially made today.

In lieu of an outhouse, a Sears and Roebuck catalog was nailed to the backside of my father's blacksmith shop and a short-handled shovel was provided to bury or throw the results of the mission over the fence into the pasture. There were two things I hated most about that situation: The sheep that would stand in the pasture on the other side of the fence and watch, and cold weather was especially miserable.

Light at night was had by lighting kerosene lamps. Even today when I smell kerosene, I am reminded of those lamps. Just before we moved from that house, we got electricity and had a drop-cord light in both the kitchen and living room. The first electrical appliance mother got was an iron and the second was a refrigerator. A double socket was put in the drop-cord socket to allow for a lightbulb and a place to plug in an appliance. I can still envision my mother standing by her ironing board, moving the iron back and forth, and the drop-cord with the double socket swinging back and forth in sync with the iron motion.

The church we attended was not held in a building but under a large oak tree in a farmer's pasture with a brush fence around it to keep the cows out. Some called it a "brush arbor." There were seats made by cutting sections from logs and standing them on end. Mostly I remember the all-day singings with covered-dish dinner on the grounds.

Early in life I also remember going to the gin with my father. Cotton was hauled to the gin in a mule-drawn wagon with high side boards because it took about fourteen hundred pounds of cotton to make a bale before removal of the seeds. On the way to the gin, I got to ride on top of the soft cotton behind my father, who sat on a board seat on the front of the wagon. Since the ride took almost an hour, I was usually asleep by the time we arrived.

What I liked most about going to the gin was when my father got paid, he would buy a big bag of jelly beans, and I would get to eat some on the way back home.

Getting candy was a rare treat during the Depression because of the cost. That's not to say that we didn't have other sweets to eat; Mother made lots of cookies and cakes using sugar and molasses, but candy was a treat for special occasions like Christmas or the ginning of the first bale of cotton. We had jellies and jams mostly made from wild blackberries and plums and some domestic fruits obtained from other farmers, usually in a swap deal.

Farm Hands

Before the Great Depression, many farmers owned their farms but ended up having to sell to people of means to survive. Many of those farmers that were forced to sell became tenant farmers. A farmer could effectively work up to twenty acres alone but needed help to work more than that.

Some families were quite large, and children that were of age served as field hands. As a matter of fact, many farming couples had as many children as possible just for that purpose. If a farmer had the acreage, but did not have a lot of children to help work it, then he had to hire help. In the South that extra help was usually African-Americans. Many white families helped each other when the farm work was heavy in a labor swap deal.

My father had enough acres to require additional field hands, especially during planting and harvesting seasons. My sisters and I were too young to be of much help on the farm. Since my father did not have a lot of money with which to pay Roy and Mary Lizzie Johnson and Humphrey, he had to pay them with produce from the garden and from the farm. This included dried peas and beans, and sweet and Irish potatoes. They were given one-third of all the canned goods from the garden and when hogs were killed, they were given one-third of all the cuts of the meat. Even then, it was not uncommon for Roy or Mary Lizzie to show up at the door in the dead of winter wanting to "borrow" a ham or a middling of meat. Naturally, my mother and father knew that the "borrowing" was, more or less, a permanent situation. Some farmers which were not as generous as my parents toward their hired help often had them stealing from them.

Age Five and Six

By the end of the fall of 1941, my father had taught me how to use the cotton scales in weighing cotton for the hired hands and record the weights in a log book. He had taken a job at Reynolds Metals Company, who was hiring a lot of people to produce aluminum for the World War II effort, and was not at home for most of the day. When I was not weighing cotton, I was picking it myself. During the season of 1941, I picked fourteen hundred pounds, which was a full bale before ginning. By the time I was a teenager, I could pick as much as three hundred pounds of cotton a day.

During the winter of 1941 and 1942, my father purchased a 1938 Chevrolet to drive to work, and we moved from the tenant house to another house about a half mile south of Tuscumbia, just off of Old Highway 43. He could no longer work the tenant farm and hold a full-time job at Reynolds Metals Company.

The move was rather upsetting to me because a lot of the things familiar from birth were lost in the move: the house with the dog trot, the creek bottom with the pond and old bee tree, the big garden, and hired hands Mary Lizzie, Roy, and Ump.

The new house had four connecting rooms, a front and back porch, and an actual outhouse. There was also a chicken house and a barn for the animals.

The anxiety of the move did not last long, because I started thinking about starting to school that fall and about riding the bus. Before I started to school, Mother began gathering up the things I would need as a first grader. Among these was a half-gallon empty lard bucket, in which to carry a lunch. The lard bucket was for warming biscuit sandwiches on top of coal-burning heaters in the classrooms. The only lunch item not taken to school was milk, which was delivered fresh daily to the school from a local dairy. There were pencils, paper, and crayons to obtain. Crayons were not rationed because of the World War II effort but were still often hard to find. Many items were rationed for the war effort: gasoline, car tires, bananas, and even pencil erasers.

Living in the new house made it easier for us to catch the bus to Colbert Heights Elementary School. On July 10, 1942, I turned six years old and started to school that fall.

The first day at school was rather upsetting for me. Just hours after arriving in the classroom and learning the teacher's name, a health department nurse arrived and had all of us students line up against a wall. She then proceeded to check each of us for head lice and started spraying a foul chemical on those where she found nits. As she worked her way down the line of children, I had become very upset because it was customary in my family to have well-groomed hair. My parents had taught me that unkempt hair was a bad reflection on the parent's effort to raise the child. If it had not been for my teacher, I would have bolted out the door. After reassurance by my teacher, she convinced me that my parents would understand if my hair was slightly messed up.

On the first day of school, my teacher, Mrs. Box, asked each student how far they could count and if they could recite the ABCs. I assured her I could say my ABCs and did. I then told her that I could count to over one hundred, and she asked me to show her. When I asked her to get the cotton scales out and I would show her, she said that was okay, but not what she wanted me to learn.

That first day was also physically traumatic for me. During recess, I was walking on a sidewalk, and a boy ran up behind me and shoved me down. I hit the sidewalk headfirst, breaking my nose, skinning my elbows and knees, and most of all, hurting my pride. Needless to say, my parents were rather upset when I got home from school that day with torn clothes, messy hair, and a face that was so swollen that I could hardly see. After that day, I had an appreciation of how hard a sidewalk could be. I am still reminded of that day every time I look into a mirror and observe my slightly crooked nose.

In November of 1942, we had eaten dinner and were working on our school homework when the wind began to blow and all of a sudden, hell broke loose. There was a loud wailing roar and the whole house began to shake. Some of the windows popped out and the front door swung open. My father jumped up, knocking a straight chair across the room in the process, and started trying to hold the door closed by putting his back against a wall in front of the door and his feet against the door. The wind was so strong that he could not hold it shut.

Mother hustled the children into the front bedroom and got us under and on the bed. I had never heard such a noise; it was like the roar of a waterfall and a thousand horses running through the house all at the same time. Then as quickly as it had started, it was over.

Rain began coming into the house through the ceiling in torrents. It continued to rain rather hard for a while before stopping. Without lights, it was impossible to determine exactly what had happened and the extent of the damage. We stayed in the bedroom the remainder of the night.

At daylight, we could see light through the cracks in the ceiling and we knew that the roof was gone. When we went outside, we found that the whole roof was blown away down to the ceiling joists, as well as the front and back porch. We knew at that point that our house had been hit by a tornado.

The chicken house and the outhouse were completely gone. One corner of the barn was damaged, but none of the animals were harmed. A cast iron wash pot that mother had on the back porch had blown about five hundred feet away and was found by the highway.

We were taken to our grandmother's house for the next two nights while my father and some good neighbors repaired our house. It wasn't uncommon for neighbors to pitch in and help when people were in need in those days.

In December of 1942, one week before Christmas, my sisters and I were informed that we were moving the next day to a community on the west side of Tuscumbia, called Valdosta. Doris, Dorothy, and I became very upset because we did not want to move to another school, but mostly because we didn't think Santa would be able to find our new house. Martha, our younger sister, who was three years old, was too young to really understand or care.

On the day that we moved we could not find Rex, our collie dog, which had been with the family since Doris and Dorothy were born. My father even went back to the house for several days to see if he could find him but never did. Three years later, Rex showed up at the house in Valdosta in really bad physical condition. He evidently had been searching for his family all that time. He had been bitten by a snake when we lived in the tenant house and as a result had a stiff neck. When he showed up at the house in Valdosta, it wasn't long before he got in a fight with a neighbor's German Sheppard and was killed, because he could not move his stiff neck enough to defend himself.

Age Six to Twelve

Valdosta is a small village on the west side of Tuscumbia. Valdosta means "valley of roses," and most of the village is in a valley split by the Frankfort Road. Some of the houses in the village are built up the hillsides and on the hilltops on the north side of the valley. The population of the village was around two hundred and fifty to three hundred, with lots of children. All of the adults obviously felt a responsibility for the children, because it was very difficult to do anything on the shady side without getting caught.

The move to Valdosta actually put us within a half mile of three aunts, two uncles, and a whole bunch of my first cousins. In addition, we were about a quarter mile from my Jackson grandparents. Moving to Valdosta was not as disturbing as we thought it was going to be. Santa found our new house and brought us the toys we were expecting.

The house was larger than the one that the tornado had hit but had only four rooms. It was better built than the two previous houses we had lived in but was still an old house in dire need of refurbishing. Instead of a fireplace, we now had a coal-burning heater in the living room and a wood-burning cookstove in the kitchen.

After the holidays, Doris, Dorothy, and I started to Valdosta Elementary School and soon made lifelong friends. Since we moved in the middle of the school year, there was some minor adjustment in my course study, but Miss Inez Walker, who taught both first and second grade, made the transition easy.

Over the next six years, I completed all six grades at that school. Miss Laverne Walker, Miss Inez's sister, taught in the third and fourth grades, and Mrs. Helen Figures, the principal, taught the fifth and sixth grades. The cafeteria cook, Mrs. Tally, cooked excellent meals, which were willingly received when I was allowed to buy my lunch.

We had the option of taking our lunch from home or buying lunch at school. I usually took my lunch but was sometimes ashamed of my sausage and biscuits, country ham and biscuits, and jelly and biscuits. Most of the other students who brought their lunches had white loaf bread sandwiches. Today I would be very proud to carry a country ham biscuit, because most would be envious of me.

I can still smell the oil that the school's floors were mopped with on a regular basis. I suppose it was used to keep down dust, but in hindsight, it was a terrible fire hazard because in each classroom was a coal-fired heater, which was often glowing red-hot.

In school, I met a boy named Jimmie (Jim) Moore, and we became close friends and have maintained contact with each other to this day. He resides in Massachusetts, where he retired from the ministry. He is one of the finest artists and sculptors I know.

Jim's mother was gravely ill when I first met him and passed away when we were in the third grade. After her death, Jim was taken care of by an aunt, grandmother, and his dad. During the summer months, Jim spent most of the daytime hours at our house and he became as close as a brother. Over the years we had many adventures together. Jim was the person responsible for giving me the nickname "Tack." As I recall, one day at school, he called me Jack, short for Jackson, and some of the other children overhearing him misunderstood, thinking he said "Tack," and that name stuck.

There was one boy in my class that was older and therefore bigger than anyone else. His name was Ralph (his last name withheld for fear that he may someday learn to read and look me up), and he was a terrible bully. Almost every day during recess, he would single out another boy, always smaller, and start picking on him by shoving him around. By the time recess was over, the smaller boy was dirty and sometimes had a bloody nose and busted lips. When Ms. Figures saw what had happened, all she would do was make Ralph sit in the coatroom for the rest of the day. I don't recall her ever paddling him as she did other misbehaving children. I honestly believe that she was scared of Ralph, because he was about as big as she was.

Being on the receiving end of Ralph's wrath, Jim, a boy named Edward Lee, and I, teamed up one day during recess and took on the task of making things right. Edward Lee and I held Ralph's arms and legs while Jim had a "come to Jesus meeting" with Ralph's face. That meeting with Ralph took place on the edge of a big coal pile out of view of the classroom windows so the teachers could not see what happened. After recess, Ms. Figures could plainly see that the four of us had been in the coal pile, but after seeing the condition of Ralph's face, she didn't ask any questions. She took Ralph to the kitchen where she cleaned his battered face and busted lips and then returned to the classroom. Needless to say, Ralph never bothered anyone after that encounter. As I remember, he was still in the same grade when the rest of us moved on to high school. I didn't know it then, but I now believe that his stairs didn't run all the way to the top floor.

The years between 1939 and 1942 had been fairly prosperous for my parents. When my dad started at Reynolds Metals in 1939, his pay was about fifty dollars per month. This was significantly more that what he was previously earning as a tenant farmer and measuring cotton allotments for the US Department of Agriculture at a rate of fifty cents per day. Even at fifty dollars a month, he was able to buy a 1938 Chevrolet car. During 1942, my father was promoted to shift foreman, a salaried position, and was making one hundred and fifty dollars per month. As the war progressed, my father worked a lot of overtime that further increased the amount of his income because there was a great demand for the aluminum that Reynolds Metals produced for the construction of airplanes.

Just before the war was over in 1945, my father started working to upgrade our house. He put sheetrock on the walls and ceiling, right over the existing boards. He then added hardwood flooring over the existing pine flooring. The finishing work on the walls, ceilings, and floors was accomplished by all members of the family.

After the war was over, my father hired a carpenter to add three rooms on to the house: an additional bedroom, a new kitchen, and a bathroom. In addition, he had a gas floor furnace installed to replace the coal-burning heater we had been using since we moved in. We thought we were really uptown with those improvements.

The Valdosta house was situated on about two acres of land, allowing us to have a large garden, an orchard, a vineyard, a barn, two cows, a hog lot, a brooder house, and pens for raising chickens, quail, and ring-necked pheasants. My father saw that I was involved in all of the work associated with those activities.

Plowing the garden was done with a walk-behind tractor instead of a mule-drawn plow. I did most of the heavy garden work because my father worked swing shifts at Reynolds and was sleeping a lot during daylight hours.

Every morning after milking, our two cows had to be taken to a leased pasture and staked out to graze on grass during day. After school, the cows were brought back to the barn, fed, and prepared for evening milking.

Stove wood had to be cut and brought into the house almost daily during the first few years in the Valdosta house. Later, after the additions to the house, my father bought an electric stove for mother, making her life, and mine, considerably easier.

Every year, the state conservation agent would borrow a pair of pheasants and quail from my father to put on exhibit in the North Alabama state fair in Florence, Alabama. My father was so proud that the exhibits were always the first stop when we went to the fair.

There was little time for Jim and me to do a lot of other things, but we usually found a way to do some. We explored every wooded area for miles around during those years. We found watermelon patches no matter how hard people tried to hide them. Whoops! I almost slipped up there.

We explored caves in the bluffs overlooking the Tennessee River and during summer months swam nude in the Tennessee, mostly on Sunday afternoons. We got caught nude a couple of times when we would be in the river and a boat with women aboard would come close to us. We often started the swimming season in March, during the AEA holidays, and would always find that the water was still pretty cold at that time of year. That water got even colder when we were in it in the nude and a boatload of people, including women, hung around to chat for what seemed forever. We were thankful that the water was murky enough to hide our nakedness. In the area where we swam, we would often find Indian artifacts like flint arrowheads and pieces of pots imbedded in the dirt banks and among the rubbish along the water edge.

It was during these years that I became intrigued with things of a technical nature. I suppose what really got me started along that line was when my dad told me how the reduction cells worked at Reynolds Metals. He told me how they cast carbon anodes for the cells and how they mixed a catalyst with the Bauxite (aluminum oxide or ore) and the large amounts of electrical energy that is required to break the molecular bond between the aluminum and oxygen of the aluminum oxide.

I got in trouble several times because of my curiosity about technology. One time I was melting and mixing different colors of wax crayons in one of mother's best pans to see what colors I could create. I found that the wax crayons were not easy to remove from the pan after melting. I think mother finally used the pan to feed our dog.

After studying about the Civil War in school and how the South mined salt peter (potassium nitrate) for one of the ingredients for gunpowder, I became interested in gunpowder. I found the formula for gunpowder in a book and knew that I was onto something. I bought some salt peter at the drug store and got some of my dad's sulfur he used to make a bug spray for his fruit trees and some charcoal from burning some wood. Mixing equal parts of each of those three ingredients and crushing them together in a bowl, I made gunpowder. I tried burning some of the gunpowder on a board, but it fizzled more that flash burned. I had read that most gunpowder was granular instead of powder form, so I moistened it with water and made a cake out of the pasty mix. After it had dried for over a week in a hot garage, I broke it up into a grainy form. I tried burning some but got the same results as before. I put the remainder of the powder in a fruit jar and stuck it on a shelf in the garage.

I had almost forgotten about the gunpowder, but one day I was in the garage and saw the jar. I got it out and went to the sidewalk by the front porch. My dad was sitting on the porch when I poured the quart of gunpowder out on a sheet of newspaper on the sidewalk. He saw what I was about to do and cautioned me about lighting the whole quart. I really didn't think that anything was going to happen except another fizzle type burn. Boy! Was I ever mistaken? When I stuck a match to the pile of gunpowder, it went off in a *flash*, there was a big fireball that went *swoosh*, and I was in it. The result was the burning of my eyebrows and lashes and singeing the front third of my hair. Evidently, the powder had dried further while in the hot garage, allowing it to flash off.

I did make more gunpowder after that, but I didn't light it in a pile. I fabricated a cannon from a one inch galvanized water pipe and threaded one end for a cap with a small hole drilled in it. I then made a base for the cannon from a block of wood and fastened it to the block with U-bolts. I put a firecracker fuse in the hole in the cap and then some gunpowder into the bore, packed some wadding in, put in a large marble, followed by more wadding, and packed that with a rod. I set the cannon on the ground at the base of a large oak tree and pointed it at an old vacant barn in some woods, about a mile from home. I got on the other side of the tree and, reaching around, lit the fuse. Almost instantly there was a loud *boom*. The cannon bucked, and the marble shot through one side of the barn and out the other side, splintering boards in the process. I just left the cannon sitting there on the ground, thinking that it was way too dangerous to fool around with any more. That ended my research with gunpowder.

Age Thirteen to Seventeen

I started attending old Deshler High School in Tuscumbia at the age of thirteen. The old school was built on property given to the city of Tuscumbia by Major David Deshler in memory of his son, Brig. General James H. Deshler, C.S.A., killed leading a charge at the battle of Chickamauga on September 20, 1863.

This was the first time that I had to ride a school bus since beginning the first grade at Colbert Height Elementary School. I could walk to school at Valdosta Elementary School.

Tuscumbia was in the process of building a new high school near the football stadium with the old Winston Home as part of the new school. I started in the seventh grade at the old high school on Main Street in Tuscumbia and at midyear moved to the Winston Home when refurbishment had been completed.

The Winston Home was started by Clark T. Barton in 1824 but was purchased by William Winston and he completed the construction of the Georgian structure in 1833. It was occupied by Union forces during the Civil War occupation. William and Mary Cooper Winston's son was governor of Alabama from 1853 to 1857.

Old Deshler High School

My class was the last freshman class to attend the old high school and the first to attend school in the Winston Home that was part of the new Deshler High School. I went through the seventh and eighth grade in the Winston Home and moved into the new high school beginning the ninth grade.

The school did not have a cafeteria so for lunch, the students had to bring their lunch, go home, or go to a restaurant downtown, all within an hour. Mostly I took a lunch and ate it in one of the school classrooms.

Sometimes, if I had the money, I would walk to Pete's Café and have a hamburger with fries and a Coke for twenty-five cents.

Winston Home

Two of my teachers at the Winston Home were Mrs. Earline Cook and Mrs. Elizabeth Ann Carmichael. Mrs. Cook was the wife of an Alabama highway patrolman, later to be elected sheriff of Colbert County. When he was shot and killed by a bootlegger, she was appointed by the governor to complete his term. She went on to be elected sheriff for several terms. While she was my teacher, she carried a forty-five caliber Derringer in her purse. She was one tough lady.

One day in class, a red wasp, which are known to be violently aggressive, lit on her forehead and proceeded to crawl down to the tip of her nose. She kept right on talking to the class without flitching, as if nothing was happening. We all sat there mesmerized by the wasp, not hearing a word she was saying, wondering what was going to happen next. When the wasp decided to fly away, Mrs. Cook swatted him from the air in one swift motion of her hand and mashed him with her foot.

Ms. Carmichael went on to teach until 1952, at which time she became the wife of US Senator Howell Heflin of Alabama.

It was during those years that I was allowed to walk to town without my parents. On some Saturdays, Jim and I would walk to town, get our hair cut at Russell's Barber Shop, go see a western movie at the Strand Theater, and then watch the girls walk by in front of the Palace Drug Store. I could do all of that for a grand total of fifty cents.

It was during these high school years that I found out that I was not the only one with a nickname. Also in school, there was a Tater Bug, Dough Belly, Pea Head, Toilet Head, and Red, plus many others, with some unmentionable.

In Valdosta, some of us guys would hang out on the porch at a store near the church we attended. The store was owned by an elderly couple, Mr. and Mrs. Hester. They had two grandsons close to my age, Roger and Bubba, who lived nearby. One day I went to the store to hang out and was promptly told to leave by Mrs. Hester. She said that I was not welcome at the store anymore but wouldn't tell me why, except that it was not good for business. About fifty years later, at a concert at our church in Decatur, Alabama, Bubba came up to me and introduced himself. I had not seen him since I was in high school. He told me that he owed me an apology for what had happened on that day at his grandparent's store; he had been stealing cakes and candy from them but had accused me of the deed when they missed the merchandise. I accepted Bubba's apology, but I still think about it often because the Hesters went to their graves thinking I had been stealing from them.

Another memorable event from high school was when two girls got in a fight over the same boy. As I remember, the girls, Maggie and Marie, were trying to date the same football "stud" who wasn't interested in either of them. That didn't make any difference to them. They went at each other scratching and clawing like two mad cats at the others' face or any exposed flesh. After a very short period of rolling and tumbling around on the ground, screaming, clawing, and scratching, a teacher intervened and broke up the fight. I don't believe I have ever seen any two bloodier people in my life. One of them had lost a bra in the process and had been clawed in that strategic area, as well as on the face and arms. Both were expelled for some time before returning to school. Soon after, both left the school for good and did not finish there.

Another event, which I claim responsibility for, goes something like this: My homeroom was in one end of the school library and the homeroom teacher and librarian was Ms. Janie B. Morgan. Ms. Morgan was somewhat up in years, as she had taught my father English when he was a student there. She was a rather large woman, in more ways than one. One day during homeroom, Jim and I were sitting near each other and I casually made the remark that "Ms. Morgan is so large that she needs two foot tubs in lieu of a bra." What I didn't know was that she was behind a bookshelf within earshot. She promptly walked over and told me that I could stay after school because of my transgressions, which I did. About a half-hour after school, I was sitting in the library in regret and feeling sorry for myself when Ms. Morgan walked over to me and said, "Mr. Jackson, I taught your father when he was in school here, and you are a chip off the old block. I may actually need to use two foot tubs, but you can't say that. Now get up, go home, and behave." I often wondered just what my father done to prompt her to say what she did.

Chemistry class was one of my favorites during high school, because we were required to do several projects during the year. One of my assignments was to make home brew (beer). The teacher wouldn't let Jim and I taste it when it was completely fermented, and we were sorely disappointed, but we had some white mice in the laboratory that really enjoyed it. After drinking the brew, the mice tried to climb the window shades, with no success. In the end, the mice couldn't walk, but we got an A plus on the project.

Another class I enjoyed was economics, taught by a gentleman named Paul Garner. Mr. Garner was a strange bird, even by 1950s standards. He often stood in front of the class with one foot against the wall underneath the blackboard and a lit cigarette hanging in the corner of his mouth. He would stand there and talk and talk, and we would sit there, watching his cigarette ashes get longer and longer, anticipating when they would fall down the front of his shirt. He would usually start his class by saying, "If you want to smoke and got 'em, light up," although no one ever did, except him. When he would turn around, he always had chalk dust all over the back of his coat, which he always wore with a white shirt and tie. Even though he was rather strange, I learned a lot about economics from him. He later left the teaching profession and went to work for the Governor George Wallace administration in Montgomery, Alabama.

It was during the summer months while I was in high school that I worked a lot for Mr. Leonard Bendall, a local farmer. He had two sons, one older than me and one younger. The older son was Hollis, who later in his life would become mayor of Tuscumbia. The younger son, Kenneth, went on to retire from Kroger Stores as a butcher. Many summers, Kenneth and I would chop cotton and corn from daylight to dark to get weeds and grass out. Those days were before the development of herbicides and agricultural poisons.

In late summer and fall, I helped bale hay and store it in barn lofts, pick cotton, and gather corn. I remember one fall we were pulling corn, and cocklebururs were as thick as hair on a dog's back and shoulder high. I had a wool jacket on, and it just attracted the burrs, which would stick tight. The tails of the mules pulling the wagon would be unrecognizable they collected so many burrs. Let me tell you, it hurts when you get swatted in the face with a mule tail full of cockleburs.

I worked for Mr. Bendall until I started to college. The last year I worked for him, I drove his truck, weighed cotton, and helped Kenneth pay the field hands.

In Valdosta, Mr. and Mrs. Sam Thompson were our next-door neighbors. Their daughter and grandchildren, James Milton and Freda Hanna, lived with them. James Milton was five years older than me. James was always hunting and trapping and could shoot a .22 rifle really well. He would sit on the back porch and drive nails, which he had barely started into a board, propped up against a shed in the backyard, by shooting the nail heads with the rifle. I really looked up to that guy.

By the time I had finished grade school, James Milton had quit school and joined the air force. He went on to retire from the air force, settled down in Delaware, received a master's degree, and became a much-published author under the name of Milton Hannah.

In the village of Valdosta, Halloween was an important time of the year for many of us. There was a group of us boys and girls that went to Valdosta Baptist Church that would go trick-or-treating together. Sometimes the boys would go their way, creating chaos where they could, and the girls would go their way doing the same, but we would always conclude our evening at Dennis Taylor's home. Grace and Dennis Taylor were a couple that never had any children of their own but on Halloween, they adopted the children of the community. They would prepare snacks for whoever would show up at their home and had games to play for entertainment. We always looked forward to going there, even into our teen years.

One person that had great influence me at that time in my life was a man that I first met at Valdosta Baptist Church. He moved into the community when the Tennessee Valley Authority (TVA) started construction of the Colbert Steam Plant, west of Tuscumbia, as the construction safety officer. I can't remember his first name, but his last name was Campbell. He was originally from Mount Hope, Alabama, and knew all about the Bankhead National Forest. When he joined the church, he became my Sunday school teacher. There were only five of us in the class, and he really took us under his wing, so to speak. He was always planning some kind of outdoor activity to keep us busy and out of trouble. Inevitably, he had a lesson he would bring out during the activities.

Mr. Campbell would carry us in his old Studebaker pickup into Bankhead Forest for a weeklong camping trip at an abandoned Great Depression era, Civilian Conservation Corps (CCC) camp that had several cabins still standing and in habitable condition. In the forest, we would take long hikes and swim beneath the Kinlock Falls on the Sipsey River.

The only drawback to camping at the old CCC camp was that there were no restrooms. To get relief meant going into the woods. I vividly remember one occasion where I went into a pine tree grove and was taking care of business when all of a sudden, yellow jackets came swarming out of the pine straw from underneath me. Before I could get my pants up, those yellow jackets had stung me several times in a strategic location. I didn't sleep very well that night. I don't know what was worse: those yellow jackets, or the chiggers that would imbed themselves in the warmest places on your body, causing severe itching.

In hindsight, I realize now that Mr. Campbell was a fantastic mentor when it came to teaching us about life's values.

College

There are many reasons why people go to college. Some go just because their parents demand it. Some go to be with friends and to just to get away from parents and play. Others end up with a BAL (blood alcohol level) higher than their GPA (grade point average) and are soon back at home until they are kicked out of the house to be on their own. Some go to just find people to date or to find their life mate. Some actually have a lot of drive and ambition and a desire to learn and make it on their own.

As for me, I believe I fell in the latter group. I saw firsthand how hard my parents had to work as tenant farmers. I too helped with the farm work from an early age. I continued doing farm work until I completed high school. As a result, I was determined to have a better life. In addition, I always had a desire to know why things in the world were the way they were.

In August of 1954, I enrolled in Florence State College, now The University of North Alabama. I planned a double major in biology and chemistry. I still lived at home and had to commute each day to attend classes. My older sister, Doris, had taken a job in Florence and would allow me to drive her car on to school after dropping her off at her office. If it was not feasible for me to drive her car on to the college, she would drop me off and I would "thumb" a ride from Florence to Tuscumbia. If I could not get a ride from Tuscumbia to home, I would walk.

One day I had "thumbed" my way from Florence to Tuscumbia and had definite plans to catch a ride with my father, who would be on his way home from work. I stood on a street corner that I knew he would pass on his way through town. I had been standing there only a short time when I saw him approaching. I stuck my hand out in a "thumbing mode," only to have him throw up his hand in recognition. What I underestimated was that he was talking to a fellow who rode with him and was preoccupied in thought. He drove right by me and went on home. Boy, was I ever let down. When I arrived home almost an hour later, I asked him why he did not stop. He said that he didn't remember seeing me. I knew then that he was certainly preoccupied.

During college, the Korean War was ongoing, and the only way I could stay exempt from the military draft was to enroll in the Reserve Officer Training Corps (ROTC). If you dropped out of ROTC, your name automatically went to the top of the draft list, so I stayed in ROTC for the duration of my college career. This allowed me to receive a commission as a second lieutenant in the U.S. Army upon graduation. Between my junior and senior years, I was required to attend a summer camp at Fort Benning, Georgia. That was a real wake up call for a country boy.

Enrolling in advanced ROTC allowed me to avoid being drafted into active army duty and provided some monthly income, which was sorely needed. I also had a part-time job in the produce department at Fuller's Super Market in Tuscumbia for additional income.

When we were sophomores on Thanksgiving break, Jim got a car from his dad and picked me up at my home. We drove west of Tuscumbia to an area called Red Rock, a rural mountainous area. We were just out riding around and viewing the fall colors when we saw a cliff across this valley. Jim suggested that we park the car and hike across the valley and climb to the top of the cliff. He pulled the car to the side of the road, and we started hiking through the woods toward the cliff. Actually, it was further than we thought, and we had to go some distance past the point of the ridge and then climb up, since it was so steep. Then we had to walk along the top of the ridge to the point of the cliff face. When we got there we were exhausted, so we sat down on the very point of the ridge and looked down into the valley below. All we could see from that point was the colorful trees and the surrounding mountains. We sat there for quite some time admiring the magnificence of God's creation and talking about how things that we take for granted are part of a grander plan. We started talking about how beautiful the view was and how it was possible that it could be that way. I told Jim about studying how the green chlorophyll of summer turns into the many colors of fall just because of shorter days, less sunshine, and cooler temperatures.

It was later that same year that Jim picked me up in a MG Convertible Roadster, which his brother Charles had driven from California where he worked as a freelance photographer. The MG was known for holding to the road on curves, so Jim suggested that we go to the TVA reservation between Muscle Shoals and Florence and run the road that looped down by the river. We were familiar with the road and knew how curvy it was. However, we did not know that a road crew had been patching some spots and left loose limestone gravel on the hardtop. We started out fine on the first few curves. Then Jim began speeding up to hit the curve that overlooked the Tennessee River. That was a grave mistake, because we hit the loose gravel, lost control, and went into a spin. We hit the metal guardrail on the riverside of the road, sheared off six metal posts, destroyed the railing, and finally stopped atop a post and bent metal rail.

We got out of the car and started surveying the damage. Jim had a bruised left elbow where he had it on the top of the little door of the MG. I had a wad of grass and dirt in my back pocket and both of us were very dirty.

MG Roadster

The car didn't fare as well as we did. There were no fenders, doors, or wheels left on the car. The steering gear, drive shaft, and rear axle were missing. All of the parts that had been torn from the car were not to be found.

We hiked back to the main highway, called Jim's brother, and then walked back to the wreck site. Charles arrived at the site at the same time as the TVA security personnel who he had called. Charles sure wasn't very happy when he saw the condition of his prized car, especially since it was a total loss.

The next morning we returned to the site and were shocked to realize that had it not been for the guard railing, we would have gone into the Tennessee River about fifty feet below. The parts torn from the car had gone into the river. We realized that we were extremely lucky to be alive.

Charles Moore returned to California and, over several years, became a world-renowned freelance photographer, selling his work to well-known magazines and newspapers worldwide. As the lone photographer on the scene at Martin Luther King, Jr.'s arrest in 1958, he captured the violence and emotion inherent in the scene. He took some of the most enduring shots of the Civil Rights Movement: the Selma bridge march, the riots at Ole Miss, and the tragedy of the fire hoses and dogs used in Kelly Park in Birmingham. He went on to win a Pulitzer Prize and write a book entitled *Powerful Days* about the civil rights movement and his life. For much of his career he worked for *Life Magazine.*

Since I started a double major curriculum, I took a lot of technical courses. In some of my advanced biology courses, we discussed the possible evolution of all living things and examined a lot of evidence that supports evolution, although there are many missing links. We learned how some things function, like a simple leaf, in support of the parent tree. A tree leaf takes water, minerals from the soil and sunlight and then, through the use of chlorophyll, formulates sugars that support a trees growth. It is not difficult to arrive at the conclusion that it is a virtual factory within itself. Understanding the complexity of the many living organisms, both plant and animal, it is hard to understand how evolution can account for it all.

While taking all of the chemistry courses, I learned that everything that we see on earth, both living and inanimate, is made up of elemental compounds in various complex combinations called matter. These elements, which make up all earthly matter, cannot be made or destroyed, but can be changed into energy, as was demonstrated by the atomic bomb. If elements such as uranium, plutonium, or heavy hydrogen are split, then heat, light, and other energy waves are the products. In making or breaking up some complex elemental molecular bonds of compounds, sometimes energy is released in the form of heat.

Much later, I learned that astronomers now say that our universe is expanding and started from one tiny point of a massive amount of energy exploding and turning into all that is in the whole universe. That explosion is known as the "big bang." The energy from the big bang started expanding in all directions and gradually started cooling. As it cooled the astronomers speculate that it turned into subatomic particles that eventually clumped together to form the elements we know today in the periodic table. They concluded from observing what they call a "red shift" or "blue shift" in the heavenly bodies that they can determine whether they are moving away or toward the observer. The biggest problem with the theory is that they cannot yet explain where all the energy for the big bang came from.

I read a book by Lee Strobel entitled *A Case for a Creator* that pointed out that several astronomers he had interviewed had determined that there is a possibility that the big bang point of energy is the Creator (God). They also determined that there are numerous astronomical parameters that have to stay in balance for this earth of ours to support life as we know it. Many of those astronomers have become skeptical about science being able to explain all that we see. One would have to conclude from what we know from science today that all of that complexity could not have taken place merely through the process of evolution. With what is known today, I have to go along with the theory that the point of energy of the big bang was God, in all of his supremacy, putting all we see in the universe into motion and then allowing some things to evolve according to his plan, as we see fossilized evidence of.

I also believe that with some tolerance, for what biblical scholars call the literal interpretation of the Bible, one can merge the creation story of Genesis with the big bang and evolution theories. One place to start is with the six days of creation not actually being twenty-four-hour days. As a matter of fact, I don't think any human can fully understand and explain what we read in the Bible and what we see as evidence of evolution. As for the belief of the inerrancy of the Bible, I do believe that God's plan for us comes through loud and clear and is inerrant in that way.

In addition to chemistry and biology courses, I took a course in religions of the world that was very interesting. The course addressed the major religions of the world, as well as some of the minor religious beliefs. I learned that, generally speaking, mankind, even to the most primitive people, is born with an innate desire to look up to or worship something or someone more supreme than themselves. The known histories of all the people that have populated the earth support that concept. Many people had multiple gods, and some only had one.

We studied the many denominations within Christianity. Each of the Christian denominations has their interpretation of the biblical scriptures and about God that separates them from the others. Some believe that God is an angry God, ready to punish anyone who slips up in just about any way. Some think God is loving and caring, ready to embrace anyone and all. Some embrace the thought that we are predestined by God and have no control over our own lives.

I don't profess to be a scholar of religions, but from my studies in college and from what I have read and observed over the many years of my life, I have developed an opinion. It seems that all religions have certain parameters that they think their God(s) exists in. It appears that all people from the most primitive to the most rich, highly educated try to keep their God(s) within certain parameters or "box in" their God(s) to fit their own lifestyles, cultures, and egos. I personally believe that the one God of the universe is so omnipotent, omnipresence, and omniscient that it impossible to place him in a box of any kind and is foolhardy to even try.

In May of 1958, I graduated from college with a double major in chemistry and biology and was commissioned a second lieutenant in the U.S. Army.

Military Service

In October of 1958, I reported for active duty at Fort Benning, Georgia. The first six weeks of my active duty I attended an officer training course, which was primarily for officers earning their commissions through ROTC. During the officer training, we started each day by doing the daily dozen exercises, followed by a five-mile run. Then, we would eat a hardy breakfast and go to training exercises for the remainder of the day. Most training days started at five in the morning and were over around five in the afternoon. Then we had time to go to the laundry, polish boots and brass buckles, etc., until bed time.

We were put through rough exercises that tested our stamina and our will to survive on a battlefield. The intent was to determine if we could function as officers and make good decisions under battlefield conditions. In some of those exercises, we had to travel at night, ten miles cross country, through swamps and deep wood, with only a compass and an M-1 Rifle. There were usually aggressor lines that we had to get through without being captured and humiliated. Some of the exercises lasted as long as thirty-six hours without sleeping.

Having grown up in the country, I fared much better than most of the city guys. I had started serious hunting at the age of twelve with my own shotgun and had run long distances most of my life. I would not hesitate to say that at least 50 percent of the city guys had never shot a rifle or shotgun before they entered ROTC summer camp.

We were required to participate in a jump tower exercise where we were first outfitted with a body harness and a static cord, like a parachutist would use. We then jumped from a short tower, which had a cable attached that descended to the ground. The static cord was snapped onto the cable and we slid down the cable to the ground.

The second part of the jump tower exercise consisted of being harnessed into an open parachute and raised off the ground by the apex of the chute to the top of a tall tower. Upon reaching the top, the parachute was released and we were allowed to fall back to the ground. Those exercises were to help us decide if we wanted to go to jump school later in our military career. I decided real quick that I didn't want to go to jump school.

After completing the officer training, we were assigned to a regular army unit for the duration of our active duty tour. I was assigned to the famous Second Infantry Division, at that time stationed in the Harmony Church Area of Fort Benning. I was given a platoon of recruits in Company C of one of the battle groups in the division, which I was responsible for training. Each platoon consisted of four rifle squads and one weapons squad. The weapons squad had machine guns and mortars in it, as well as rifles. Each squad consisted of twelve men and a squad leader. Each platoon had sixty-seven men counting the sixty-five men in the squads, the platoon sergeant, and the platoon leader.

At that time, the second division was training replacements for the third division stationed in Europe. Our division would put the recruits through basic and advanced training and then ship them out to the third division. The third would then rotate their troops back to the United States.

When I arrived in my platoon area, the first person I was introduced to was the platoon sergeant. Sergeant Hierhi was a short, stocky Hawaiian of twenty-four years of military service. He had fought in Korea and was well decorated. I could tell right off that he knew what he was doing. I told him that I was fresh out of college and the officer training course and had no experience leading a platoon of recruits. He responded by saying "Lieutenant Jackson, don't do anything which you are not sure of without talking to me, and I will take care of you. Maybe together we won't do anything stupid." In hindsight, that was a terrific way to start out, because I went through my active duty time without any major incidents and without looking like a fool.

The second person I met was Capt. Leo Anderson, the company commander, a salty old man with over thirty years in military service, including time in Europe during World War II and in Korea. He had more metals than he could pin on his chest, having been wounded in the Battle of the Bulge in World War II and again in the Korean conflict. He was single, had lost his wife to cancer some years before, and considered the army his family.

Capt. Anderson introduced me to the other four platoon leaders of the company, all second lieutenants, and the Company executive officer, a first lieutenant.

One of the other second lieutenants was a West Point graduate and let it be known right off. He asked me when I graduated from the Point and I replied, "I got my commission through ROTC."

He immediately said, "Oh…you are one of those ninety-day wonders."

You can imagine how that went down, since three of us were ROTC officers and one, Lieutenant David Davoli, had gotten his commission through ROTC after serving in Korea as an enlisted man. I thought Lieutenant Davoli, previously wounded in Korea, was going to kill the West Pointer. Lieutenant Davoli, turning red-faced with anger, pulled his shirt off, showing a scar that started on his left hip and extended across his body to his right shoulder. He had been hit by shrapnel and lay in a rice paddy for two days, holding his guts in, before being found and moved to a field hospital. It had taken him more than a year to recover from his injury before going to college and getting his commission as an officer. Davoli said, "Does this look like I'm a ninety-day wonder?" The guy from West Point turned red and almost swallowed his tongue trying to apologize.

Several months later, I was asked to take over the West Point lieutenant's platoon because he was having so many problems with the southern African-Americans that made up over half of his platoon. The African-American recruits did not really like southern lieutenants and they disliked arrogant, northern West Point graduates even more, to say the least. Their dislike showed up as many discipline problems.

After taking over the platoon with Sergeant Hierhi, things went fairly well for several weeks. Then one morning Sergeant Hierhi walked into my office and said in an excided tone, "Lieutenant Jackson, you have a problem!"

I responded, "Just what problem do I have Sergeant?"

Sergeant Hierhi related to me how some of the African-Americans were complaining about three of their own, which were not bathing on a regular basis and had gotten rather odoriferous. All they were doing, in lieu of bathing, was putting on more high-powered cologne. I asked Sergeant Hierhi what he would do to resolve the problem, and he replied, "We could get one of their own, which they respect, and make him responsible for seeing that they do what needs to be done on a regular basis."

I said, "Sergeant, you are closer to the men than I am, so pick out someone to put in that job."

The next morning, Sergeant Hierhi came into my office with one of the biggest African-Americans in the platoon. His name was Anthony Croom. He was from South Carolina and was a person that, anyone could tell, was raised up right. I told Anthony what I wanted him to do and asked if he was willing to take on the task. He said, "Sir— Can I do whatever I think is necessary to do the job?" I assured him that he could do that, and he left the room smiling, happy to have that authority.

I became concerned that I might be doing the wrong thing, so I got Anthony's file out and reviewed it just for my own assurance. That's when I found out that he was college educated, with a degree in human resources and psychology, and also came from a well-respected business-oriented family. That took some of my worry away.

That evening, I was about to leave for the day when Sergeant Hierhi came running into my office very excited.

"Lieutenant, come quick."

So I got up and followed him. We went into the barracks building where the recruits lived and into the shower room. There, standing under the shower heads were three naked recruits with Anthony Crooms standing nearby, holding an unsheathed bayonet. As we entered the shower room, I overheard Anthony say, "If you don't start showering regularly, I will cut your sorry throats." When we entered the room, Anthony glanced at me, smiled, and winked. I asked him to step outside for a minute and asked him what was going on. He said that everything was under control and to not worry, that he was not really going to use the bayonet, but that they didn't know that. I told him to be careful, and Sergeant Hierhi and I walked back to my office. I didn't have any more complaints about dirty recruits after that.

Many times after that when the platoon was out in the field getting ready for training exercises, Anthony would get up in front of the bleachers and sing *Old Man River* and more than half of the men, including me, would get teary eyed. I believe he could sing that song more soulfully and beautifully that anyone I have ever heard. That guy was a real leader and before his basic training was completed we recommended him for OTC (Officer Training Course), which he later attended.

At that time in history, it was customary for a new officer entering active duty to be invited to his superior officer's home for dinner as an ice breaker. Since Captain Anderson was not married and could not cook, he left it up to his executive officer to do that. The executive officer was a black first lieutenant, West Point graduate from Connecticut. The evening that I went to his home, he evidently had done his homework, because one of the first things he ask me was, as a southern boy, did I have any hang-ups with eating dinner with a black family? I was a little taken aback with that question, but I assured him that it did not bother me at all. I told him of Mary Lizzie, the black lady that helped mother, practically living with us when my sisters and I were young and all of us eating together. I suppose he thought that all southerners were racists, since he was from Connecticut. Everything went well that evening, with no more indications of possible animosity from the executive officer toward me.

My biggest job while on active duty was teaching recruits how to use the 81 mm mortar, 106 mm recoilless rifle, and light machine gun. I probably spent half of my time preparing lesson plans that were reviewed by the company commander and the other half actually teaching. In April of 1959, I finished my tour of active duty and returned to civilian life in Tuscumbia. I remained in the active army reserve for sixteen years, reaching the grade of captain, before resigning my commission in 1975. The summer camps and multiple drills were interfering with my work career, so I felt it best to resign my commission.

Marriage

When I started my sophomore year in college, I started going places with a boy named Bill White. Billy had a car and I didn't, so he would come by my home and pick me up from time to time. One day I was at home and my younger sister, Martha, had a school friend of hers there who was going to spend the night. Her friend's name was Judith Kent. The next morning I said something about Billy coming by to pick me up and found that Judith had a crush on him.

Several months later there was to be a New Year's Eve program at church, and Martha said that Judith wanted me to get Billy to ask her for a date for the church program. The next time I saw Billy, I told him of the plan and he promptly told me that he really didn't have any feelings for her and would not ask her out.

I had been around Judith enough when she was with my sister and thought she was a really neat girl. Since I did not have a date for the New Year's Eve church program, I told Billy that if he wasn't going to ask Judith for a date, then I was. He said, "Go ahead," and I did. Judith accepted and that was our first date of many.

You have probably heard some girls say that they will not kiss on the first date. At least that was often the way it was during my youth. Well, Judith was not one of those girls, and that is all I am going to say about that. We both had a really good time on that first date. I believe that we both fell hopelessly in love on that night and have not fallen out since. We went on to date for the remainder of time that I was in college. When I entered active military service at Fort Benning, we didn't see each other very much. In April of 1959, I came home from service, got a job, and we got married in May of that year in a Friday evening ceremony. We went on a weekend honeymoon to Lookout Mountain at Chattanooga, Tennessee, because I had not been working long enough to have accumulated any leave time.

When we were going back home from the honeymoon trip, we were talking about how much money we had and realized that we only had twenty-five dollars between us, and a fifty dollar car payment coming due.

We could not afford to rent an apartment, so we lived with Judith's parents for a month and then with my parents for a month. We then rented an upstairs apartment in Billy White's aunt's house and lived there for the better part of year, before moving into a garage apartment.

Just a little over a year after our marriage, Jana, our first child was born. The night she was born, Judith was supposed to be in the graduation ceremony at the college, but was in the labor room at Keller Hospital in Tuscumbia.

When we moved to Greeneville, Tennessee, when Jana was less than a year old, Judith had a hard time adjusting. That was really the first time that she had been away from her parents and with a child to care for, she was almost overwhelmed. Her mother came to Greeneville as often as she could to make the situation as easy as possible.

Two years after Jana was born, Kevin was born in Greeneville, and two years later, Alicia was born, also in Greeneville. When each was born, Judith's mother came and stayed with us for several weeks to assist with the children. She was a godsend, to say the least.

Each of our children went on to grow up without us killing them, getting a college education, and marrying well. All three had finished college the year that I reached forty-eight. Jana received a registered nursing (RN) degree from the University of Alabama Birmingham (UAB); Kevin got building science and civil engineering degrees from Auburn University and the University of Alabama Huntsville (UAH), respectively; and Alicia got a management information systems degree from UAH. In the same time frame, Judith got her MS in home economics. In addition, she has become an accomplished artist.

Jana and her husband, Keith, have two children, Emily A. and John W. Frost. Kevin and Dena do not have any children. Alicia and Kris have two children, twins Hannah A. and Braden K. Hahn. There is not a father or grandfather alive that is any prouder of their family than I am.

The memory of years spent with my children while they were growing up is very precious to me. During their preteen years, there were almost nightly wrestles on the floor. I vividly remember those, because the three of them could almost get the best of me.

Watching them develop through the many stages of their lives brought back memories of my own life and the differences between the phases of their life and mine. Seeing the grandchildren grow up, in like manner, brought back those memories again, and I will hold them deep within my heart for the remainder of my life.

First Jobs

Upon coming home from military service, I went to work for a company in Tuscumbia, Alabama, called Muscle Shoals Electrochemical Corp. The company manufactured electric furnace fused ceramic refractory products, primarily magnesium oxide used in the heating elements of electric stoves, hot water heaters, irons, and etc. I worked in the laboratory of the company as a chemistry and quality control technician.

After working there for almost a year, I learned that the company manager and the chief engineer were planning on pulling out and going to Greeneville, Tennessee, to build a new competing company. The three shift foremen and I were asked be part of the team in building the new company. In early fall of 1960, we left Tuscumbia and travelled to Greeneville, Tennessee, to start building the new plant in a hay field north of town.

The new company in Greeneville was named Tennessee Electro-minerals Corp. and manufactured other ceramic refractory products besides magnesium oxide including mullite, fosterite, and fused silica, all used primarily in fire brick. In addition, the company occasionally fused amorphous uranium for the making of nuclear fuel rods.

My job at the new company was to set up and run the laboratories for chemical analysis and quality control. Within one year, working long hours, we had the company in Greeneville up and running and manufacturing the products. We had standing orders before the first products ever came out of production.

Later, I started working in customer support, travelling at times to customer factories to troubleshoot problems with our products.

The company prospered for the first three years until the chief engineer and general plant manager, Elmer Allen, pulled out and started his own machine shop company. After this loss, the company president found it difficult to run the company. Before Elmer left he oversaw the day-to-day operations and the interfacing with the foremen and maintenance lead personnel. The president, because of his personality, had a hard time getting along with the people he had to interface with, and it soon started to show in production and the attitude of the employees. He was a brilliant man, from a technical standpoint, but didn't have a lot of common sense (Some people call it walking around sense) to go along with it.

Several key employees resigned, causing some disruption in maintaining and operating the plant. One of the original foremen, Vernon Cooksey, died of a heart attack and another, Fred Belew, was severely injured in a fall in the plant and was never the same again. The other original foreman, Hershel Muston, returned to Alabama to work for another company. In the late spring of 1964, I left the company and returned to Alabama to work. After all the departures, the company struggled for another three years and then merged with Combustion Engineering of Chattanooga, Tennessee.

We returned to Tuscumbia, Alabama, for eight months, and I worked for a pest control company and an insurance company to make ends meet. In January 1965, I began work for the National Aeronautics and Space Administration (NASA).

NASA MSFC

I started to work for the National Aeronautics and Space Administration (NASA) at Marshall Space Flight Center (MSFC) in grade GS-9 as a laboratory technician in the surface treatment and electroplating facility in the manufacturing engineering laboratory. My section chief and immediate supervisor, Max Sharpe, was over that facility as well as the precision cleaning facilities, and paint shop.

When I began work there, it was the early days of NASA and the Von Braun team members were in the highest supervisory positions at MSFC down to some of the division levels. My division chief was Otto Eisenhardt, and the laboratory director was Werner Kuers. They were all part of the Von Braun team brought to the US from Germany after the end of World War II.

It was my responsibility to pull samples daily from each of the many chemical tanks and run a complete analysis on the solutions. After the analysis, if any additions to the solutions were required to bring them up to specifications, it was my job to do so. I had been working for MSFC for about six months when Max was promoted to chief of the Manufacturing Engineering Branch, and I was moved to his branch and reclassified as an aerospace engineer.

I had been at NASA for over a year when one of the sheet metal workers walked up to me as I was walking through the shop. He said, "Can I ask you something?"

"Okay!" I responded.

He then asked, "When you started working here at NASA, we had been told that NASA hiring was frozen. And then all of a sudden, you show up as a laboratory technician. What we want to know is, is that your only purpose for being here?"

"I don't know what you mean." I replied.

He looked at me real hard for what seemed like five minutes and then began to speak, "Will you tell me the truth if I ask you another question?"

"Sure—if it's not something personal," I said.

He then replied, "Now don't get mad at me. Are you a CID (Criminal Investigating Division of the FBI) agent planted to watch what goes on here?" After I quit laughing, I assured him that I was not.

With that episode, I noticed a marked difference in the friendliness and how I was greeted when I walked through the shops.

One day I was told to report to Otto Eisenhardt's office and upon entering, he asked me if I would be interested in a management position. He said that he saw in my file that I was an army officer and he thought I could better serve the agency as a manager. I told him I would, and on October 19, 1971, I was promoted to chief of the metal finishing section with an increase in pay to grade GS-12. However, I still reported to Max, who was chief of the Manufacturing Engineering Branch.

Five years later, on October 13, 1976, I was promoted to chief, mock-up and prototype assembly branch in the grade of GS-13. In that capacity I was over the surface treatment and electroplating facility, paint shop, wood model shop, plastics shop, heat treat facility, sheet metal shop, precision cleaning facilities, and printed circuit shop.

In 1977, I moved from the Manufacturing Engineering Laboratory to the Materials and Processes Laboratory (M&P) where I was a group leader in the Chemicals and Non-Metals Processes Branch for four years.

In 1986, I went to work in the space shuttle Main Engine (SSME) Project Office and was reclassified as a liquid propulsion engineer. There I worked with Pratt Whitney in the development of a new liquid oxygen turbo-pump for the engine.

On April 22, 1987, I was promoted to chief, Chemicals and Non-metals Processes Branch of the M&P Laboratory in grade GS-14.

During my years at MSFC, I worked on the Saturn V first stage booster, the space shuttle Main Engines (SSMEs), space shuttle Solid Rocket Boosters (SRBs), Skylab, High Energy Astronomical Observatory (HEAO), Laser Geodynamics Satellites (LAGEOS), Hubble Telescope, Lunar Roving Vehicles (LRV), MSFC spray on ablator (MSA), Spray-On Foam Insulation (SOFI) and other programs. It was during these programs that NASA started the early engineering phase of the space shuttle transportation system, twelve years before the first test flight in 1881.

The first two, first stage, Saturn V boosters were built in the manufacturing facilities at MSFC by government personnel, with the contractors observing, before being turned over to the contractors to be built in New Orleans at the Michoud Assembly Facility. We had most of the manufacturing problems resolved before we turned it over to the contractors.

My section performed the chemical milling on the Saturn V first stage booster, bulkhead gore segments, and cleaned them for welding. All other LOX (liquid oxygen) fuel tank side segments were also cleaned for welding in the surface treatment facility. All of the precision cleaning of the booster engine tubing and components was performed in the precision cleaning laboratories of the section.

After the booster LOX and fuel tanks were assembled, they were placed into a tower in Building 4707, where they were cleaned internally by my section. To clean the internal surfaces, technicians had to dress in white nylon clean room suits, including booties, and climb the slosh baffles which were installed at various levels inside the tank, wiping all of the surfaces with pharmaceutical grade alcohol as they went. At the entrance to the tank, a portable clean room was set up to keep from re-contaminating the inside of the tank. What we failed to do during the first tank cleaning was to provide proper ventilation into the tank and safety nets below the slosh baffles. Yep! You guessed right. The guys cleaning inside the tank got so drunk from breathing the fumes of the alcohol that they could hardly climb down from the slosh baffles. After this incident we provided filtered air conditioning into the tanks, to the disappointment of the workmen, as well as safety nets below the baffles.

Unless you have been inside of the LOX and fuel tanks of the Saturn V Rocket first stage, it is hard to conceive just how large it is. That beast is thirty feet in diameter! The fuel tank is large enough to hold a freight train engine.

When the first Saturn V launched, I was at home watching on TV. When the five engines started firing, it seemed that it would take forever to lift from the launchpad. I found that I was just holding my breath, hoping it would not fall over, until it started to move upward. For a country boy, it was just hard to believe that something as big as the Saturn V could ever get off this earth. When it did start moving up, cold chills went up and down my spine from knowing that I had a hand in making it happen. After that first launch, if someone within NASA had told me that we were going to launch the Empire State Building, I would probably have believed them.

Skylab

During the Skylab program, which was designed at MSFC, some of the major components were built there; components like the telescope, the telescope mount, and the ATM (Apollo Telescope Mount) cameras. In 1969, McDonnell Douglas Corp. was awarded a contract for the conversion of two existing S-IVB stages to an orbital workshop configuration. The orbital workshop was renamed "Skylab" as the result of a contest. The Skylab was actually the upper stage of the S-IVB. Weighing around one hundred tons, it was put in orbit on May 14, 1973, with a two-stage Saturn V rocket. It was subsequently visited three times by crews from 1973 to 1974 and returned to earth in 1979 as a fireball over Western Australia, leaving debris scattered over a long path.

One of the jobs that MSFC had during development and building of the Skylab was to fabricate and paint the ATM camera housings.

After machining, the aluminum housings were anodized and prepared for painting with a special paint called Micatex. The Micatex paint was formulated by a branch of the M&P Laboratory using a commercially available latex paint. The branch personnel would add finely ground mica and white titanium oxide to a given amount of the latex paint, which acted as the binder for the whole mixture. The paint shop in my branch had already successfully painted several housings when all of a sudden, the paint would no longer cure properly.

After two unsuccessful applications of the paint, I told my laboratory lead engineer that the paint was bad. He related that to the M&P director, Bob Schwinghamer, who promptly called my laboratory director, Dr. Matt Siebel, and told him that my personnel were ruining the paint or not applying it right.

I was called by my director, who chastised me, telling me to get things straightened out. I called Marshall King, the branch chief of the M&P Laboratory, and invited him and his people to come to our paint shop and apply the paint themselves. The next day they came to our shop and painted the camera housings using freshly mixed paint. We locked the shop, and I had Marshall hold the keys until the following morning. When we opened the shop, the paint was uncured as I expected. Marshall and his personnel went back to their laboratory, and I started making some phone calls. I found out that the commercial supplier of the latex paint had changed the ratio of latex to curing agent in the paint, which left it marginal for curing in the best of conditions. With the addition of mica and titanium oxide, the ratio was such that it had no chance to cure properly. I called Marshall and told him what I had found and then called Dr. Siebel to relay the news. Marshall King and his personnel corrected the problem and started making good paint again. None of my painters nor I ever received an apology from anyone involved in making the accusations about us ruining the paint.

Severe damage to Skylab was sustained during launch, to include loss of part of the micrometeoroid shield, sunshade, and one of two solar panels. The solar panel was hung and without it, there was a severe power deficit. If repairs were not made by the astronauts, the whole program would be lost. Without the sunshade, the plastic materials used inside the station would melt, contaminating it with poisonous gases and making it unsafe.

All of the NASA centers went to work on a crash basis to put together the cutting tools required to release and deploy the solar panel. They had to fabricate telescoping poles to deploy the solar panel and make a deployable sunshade for the one lost. All of the tools and sunshades had to be taken into orbit by one flight and utilized by the astronauts during a spacewalk to save the Skylab.

I received a phone call at two o'clock in the morning from my branch chief, after it was determined what had happened to the Skylab during launch. I was told to report to work, expect to stay indefinitely, and call all of my technicians into work to do whatever was required to support an effort to save the Skylab.

Within two hours, I had all of my technicians at MSFC and starting to process some pieces of hardware that had already been fabricated. We all remained at work for the better part of a week, sleeping on a pallet of bubble wrap in an upstairs room for short periods of two to three hours at a time before working again.

We (NASA) brought people in from New England who sewed sails for big sailboats to sew sheets of Mylar (a strong plastic) together to fabricate the deployable sunshade needed to save Skylab. When the sewing was complete, we laid the sunshade out on the floor of Building 4707 for painting. We were preparing to paint the sun shield, but did not have the right paint called out on the drawing specs. I called a friend who was the manager of a local paint store in Huntsville, Alabama, and asked him if he had the required paint. He said he did, but I knew that we didn't have time to go through the government procurement process. I asked him if we could exchange some paint we had on hand for the right kind, and he said we could if we didn't caught. I loaded several gallons of paint into a NASA truck, drove to the local paint store, and swapped the paint.

After I got back, I became concerned about possibly violations of government regulations, so I called

Dr. Siebel, the laboratory director and started to tell him what I had done. He stopped me in mid-sentence and said, "Don't tell me anything else; just do the job."

After painting the Mylar shade and allowing the paint to dry, it was taken into our largest clean room and cleaned by wiping it down with mild solvents. After drying, it was folded and packaged for flying to the crippled Skylab.

When I turned the time sheets into the payroll office, I got a telephone call from their supervisor. I was told that I could not work the employees those long periods without violating regulations. After the call, I called Dr. Siebel, who told me to not worry and that he would cover for me. All of the people at MSFC got the job accomplished during a long work week, and the Skylab was saved by the crew deploying the sunshade and cutting the solar array loose so it could be deployed.

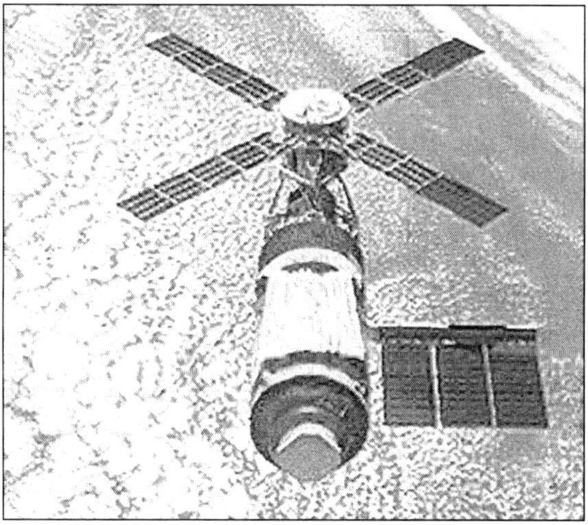

Skylab in orbit showing the White Sun Shade & stuck Solar Panel (left side) Deployed

About a month after the Skylab was rescued by the astronauts, I got a call from my laboratory director's office and was told to report to the center director's conference room. There were several other supervisors in the conference room, and none knew why we were there. Rocco Petron, the center director, who had been brought in to break up the high level managerial clique of the Von Braun Team at MSFC, walked into the room. I think we all thought that we were in deep trouble at that point. He immediately put our concerns to rest by telling us that we were to receive awards certificates.

When he called my name to come forward to receive my certificate, he looked directly at me and said, "Some of the activities that went on behind the scenes in the effort for Skylab recovery I don't particularly condone; however, I do appreciate the effort put forth to meet the tight schedule and would like to personally thank you for the effort."

I still wonder today just how much he knew about all I had done.

The lesson learned from that experience is this: Sometimes, based on the circumstances, it is better to ask forgiveness, if caught, than to ask permission.

Technologies

With all of my experience in supervising the manufacturing shops and later working in the M&P Laboratory, I was asked to participate in many Quality and M&P reviews at NASA contractor facilities for spacecraft components like the SSMEs (Space Shuttle Main Engines), SRMs (Solid Rocket Motors), ETs (External Tanks), and others.

I attended many short courses and seminars on various technical subjects. One that really impressed me was a course on the principles of electroplating and electroforming by Dr. William Blum, a world famous professor and industrialist. He was ninety-eight when I attended his course in Los Angeles, California. One of the most interesting things that he pointed out was how the best brighteners were discovered for cadmium electroplating baths. He said it was in the early 1900s when a chemist in an electroplating shop was unable to get the bath chemically balanced to produce a bright finish. He had tried and tried but failed. One morning, the chemist came to work and there on a workbench was some of the brightest cadmium plated parts he had ever seen. He started asking the electroplaters what they had done to get such good results, but no one knew anything. All they knew was that the night shift had electroplated the parts lying on the workbench.

That night the chemist got out of bed, because he could not sleep, thinking about the bright cadmium, and went to the factory. He started questioning the electroplater that usually worked the cadmium line at night, but he denied knowing anything.

Another fellow nearby had overheard the questioning and just as the chemist was about to leave the building in disgust, he cut him off. He said, "I usually don't rat on fellow workers, but I saw that guy pee in the Cadmium tank last night and after that, all of the electroplated parts coming out of the tank were bright." The chemist was dumfounded!

The next day he pulled a sample from the cadmium tank and performed a complete quantitative and qualitative analysis on it. What he found was that the urea in the pee acted as a brightener in the bath. That finding resulted in synthetic urea brighteners being used in practically all cadmium electroplating baths for many years.

After taking the course under Dr. Blum, anytime that MSFC personnel were to review contractor electroplating or electroforming processes, I was asked to participate. For example, when MSFC let the contract with Rocketdyne Corporation for the space shuttle Main Engines (SSMEs), I was asked to attend a process review.

The throat of the SSME nozzle is made by electroforming nickel over a slotted copper alloy mandrel. After letting the contract with Rocketdyne, I was asked to attend anytime there was to be Quality or M&P reviews or audits.

During my years in the M&P Laboratory, there were several engineers in the Chemicals and Nonmetals Branch that worked together to build what would later be called the Productivity Enhancement Center (PEC). It became nationally known for being on the cutting edge of technological developments in materials and processes.

Some the developments from that facility consisted of a spray on ablator (MSA) for the SRB components, a Spray-On Foam Insulation (SOFI) for the ET, a process development cell for the SOFI, and a Hydro-laser (high pressure water) process development cell for the removal of MSA from recovered SRM and SRB components, as well as others.

I received a NASA medal and certificate for development of the SOFI process cell 1986.

NASA's Jackson receives medal for exceptional engineering

Charles H. Jackson, an aerospace engineer at the Marshall Space Flight Center (MSFC) in Huntsville, Ala., was presented a NASA Exceptional Engineering Achievement Medal recently. The medal was presented by Dale D. Myers, NASA's deputy administrator, and J.R. Thompson, MSFC director, at the Center's Annual Honor Awards Ceremony.

Jackson, who works in the Materials and Processes Laboratory, has been employed at the NASA facility since 1965.

Jackson was honored for "exemplary engineering achievement and leadership in development of the Space Shuttle external tank thermal protection system, and for exceptional foresight and technical judgment in establishment of the advanced research and development facilities in use at MSFC."

He is a 1954 graduate of Deshler High School in Tuscumbia. Jackson received a bachelor of science degree in chemistry from UNA (then Florence State) in 1958.

The Tuscumbia native is the son of Fletcher Jackson of Tuscumbia. Jackson is the son-in-law of Mr. and Mrs. Wilson Kent of Hartselle.

Jackson and his wife, Judith (a 1960 Florence State graduate), have three children and reside in Hartselle.

The Marshall Space Flight Center has a leading role in the space program. Currently, the Marshall Center is responsible for a wide variety of NASA projects ranging from development of the Edwin P. Hubble Space Telescope and production of the propulsion elements of the Space Shuttle to management of Spacelab Earth-orbital missions and other payloads for the Space Shuttle.

Also, the Marshall Center has been given a substantial role in the development of Space Station, a permanent manned facility proposed by President Reagan to be in orbit by 1994.

The station would offer the capabilities of scientific research and technology development by both government and industry; the commercial use of space in such areas as the manufacture of critical materials and pharmaceuticals not available on Earth; the assembly, servicing and repair of satellites and other large structures in space; and research focused on extending a human being's staying time in space as a first step toward more ambitious manned space programs.

The University of North Alabama Alumni Paper (article)

As was the technological developments for earlier major NASA projects, the preliminary design engineering and materials and processes developments for the space shuttle began at least ten to twelve years prior to the first launch in 1981. We started looking at the technologies that would be needed for the space shuttle in the PEC as early as 1970.

The design of the space shuttle stack, ET, SRBs, and Orbiter also began years earlier and went through many reviews before being finalized. Even after all of the exhaustive preliminary design work and technological developments, there were many problems that had to be overcome during the later building of the actual hardware.

At MSFC, problems weren't problems; they were challenges and opportunities. There were many headaches and a lot of time expended overcoming those challenges, but we had the opportunity to do things that mankind had never done before and we did it. Sometimes we did things that some said were impossible, but they forgot to tell us, so we did them anyway. We had a lot of farm boys working at MSFC who were used to solving their own farm problems and who didn't know that some things were impossible, so they just got the job done.

Challenger Accident

Because I had been involved with major components of the space shuttle in 1984, I was asked to be the MSFC Science and Engineering Office alternate to Billy K. Davis on the pre-launch Ice/Debris Team. The team leader at Kennedy Space Center was Charlie Stevenson and his alternate was Greg Katnic. Each of the NASA Centers that had engineering involvement with the shuttle had a member on the team.

The team's mission was to look for any ice buildup on the shuttle ET just prior to launch that might break loose and any loose debris on the launch pad that might hit the fragile tiles on the space shuttle Orbiter during the launch. The team used binoculars and Infrared scanner guns to determine if there was ice buildup on the external tank. Over a period of three years, I alternated with Billy K. Davis on eight launches until just before the Challenger accident.

For all of the launches that I worked, I was at Launch Pad 39-A. That pad, along with 39-B, was built to support the Apollo program. Pad 39-A was modified to support the first space shuttle launch. After several launches from 39-A, modifications were initiated for Pad 39-B. It was not until the launch of 51-L (Challenger) on January 28, 1986, that the Pad 39-B was used for the first time.

Ice/Debris Team (1985)
from left Charlie Stevenson, Charles Jackson, Et Rep., Orbiter Rep., Greg Katnic

About a year before the Challenger accident, I was at Kennedy Space Center working a launch as a member of the Ice/Debris Team. I was on the retractable access arm, using an infrared gun to take some readings on the ET when all of a sudden, I heard a voice behind me say, "What are you doing? You are not going to shoot a hole in the tank, are you?"

I turned around and there stood Ellison (El) Onizuka, in his space suit, ready to get aboard the space shuttle for his first flight. I replied, "No! I am on the Ice/Debris Team, and this is an infrared gun that we use to determine if there is ice buildup on the ET."

In response, he said, "Good! Because this is my first flight, and my legs are already shaking, and I am scared to death."

He then asked to see the infrared gun and asked me how it worked and what my name was. About that time, two more astronauts came along the arm, and he told them what I was doing. They thanked me for what I was doing, and I left the arm by way of the elevator door when the boarding technicians came out, ready to assist the astronauts in the boarding process. That was the first time I met El Onizuka, but not the last. A year later, our paths would cross again.

The flight 61-C, just before the Challenger flight—51-L, was supposed to be worked by Billy K. Davis. He asked me to work that flight so he could work 51-L, because it was to be the first launch from the new Launch Pad 39-B. Flight 61-C was supposed to be launched on

January 12, 1986, but because of several technical problems, it was not launched until January 18.

We had to load and unload the fuel and oxygen four times before we finally launched. After each de-tanking of the fuel and oxygen, the Ice/Debris Team had to perform a detailed inspection of the ET and the launch pad. We had to make sure that there were no cracks in the cryogenic foam insulation on the ET. We also had to re-inspect the launch pad for any debris prior to the next tanking and then again just before the next launch time, usually while the astronauts were boarding. Between January 12 and the launch of 61-C on the eighteenth, the Ice/Debris Team did not get much sleep.

After the first two tanking and de-tanking operations, the news media started chastising NASA for not launching the shuttle in a timely manner, since it was supposed to be entering, what was termed, the operational phase of the program. In addition, the NASA Headquarters Public Affairs office started applying pressure to launch on time, because they felt the pressure from the news media.

It was between one of the tanking and de-tanking operations that I ran into El Onizuka again. During lunch I had gone to the cafeteria and bought a sandwich and returned to a conference room to eat and there El Onizuka and Judith (Judy) Resnik were in the room. Before I could introduce myself, El turned to Judy and told her who I was and how we had first met a year earlier on the launch pad access arm. He related to her how our conversation on the access arm had helped him settle his nerves somewhat before his first launch into space. The three of us sat in the conference room for the better part of an hour eating our lunch and just chatting about our careers with NASA.

Shuttle launch 51-L with the Challenger Orbiter was scheduled for January 22, 1986, just four days after the launch of 6I-C. The Challenger had flown nine times, successfully, before 51-L. Shuttle mission 51-L was much like most other missions. The Challenger was scheduled to carry some cargo, the Tracking Data Relay Satellite-2, as well as fly the Shuttle-Pointed Tool Astronomy (SPARTAN-203), a free-flying module designed to observe tail and coma from Halley's Comet with two ultraviolet spectrometers and cameras.

The most unique thing about the flight was that it was the first of a program called Teachers in Space Program (TISP), and Challenger was scheduled to carry Sharon Christa McAuliffe, the first teacher in space. She had been selected from over eleven thousand applicants and given the training needed for the flight. She was excited to be selected and was quoted as saying, "I watched the Space Age being born and I would like to participate."

Because Christa was to be the first teacher in space, and the launch being in January, we were informed that President Reagan wanted to talk about it in his State of the Union address. Almost everybody within NASA felt it was very important to get the launch off in a timely manner because of the president's interest and the heat from the news media during the previous launch.

Along with Christa, the rest of the crew was Commander Francis R. Scobee; Pilot Michael J. Smith; Mission Specialists Ronald E. McNair, Ellison S. Onizuka, and Judith A. Resnik; and Payload Specialist Gregory B. Jarvis.

Mission 51-L was plagued by problems from the very beginning. Launch, initially scheduled for January 22, slipped to January 23, then to January 24 because of delays in launching 61-C. It was then reset for January 25, because of bad weather in Dakar, Senegal, an abort-landing site. The launch was again rescheduled for January 27, because of weather and turnaround time requirements. Then it was delayed another twenty-four hours because a ground service equipment hatch-closing fixture could not be removed from the Orbiter hatch. That problem was fixed, but then cross winds exceeded launch limits. Challenger finally lifted off at 11:38 a.m. Eastern Standard Time, and...

On January 28, 1986, seventy-four seconds into the mission, the Challenger (Flight STS 51-L) exploded, killing the crew of seven and initiating worldwide mourning.

People were stunned worldwide. The crowds of observers, family, and friends at Kennedy Space Center stood around stunned and speechless. Families all over the world tuned in to watch the news.

Later in the day, the president made a speech to the nation, saying, "Today is a day for mourning and remembering. Nancy and I are pained to the core over the tragedy of the Challenger. We know we share this pain with all of the people of our country. This is truly a national loss. Nineteen years ago, almost to the day, we lost three astronauts in a terrible accident on the ground. But we've never lost an astronaut in flight. We've never had a tragedy like this. And perhaps we've forgotten the courage it took for the crew of the shuttle. But the Challenger Seven were aware of the dangers and overcame them and did their job brilliantly. We mourn seven heroes."

After President Reagan made his speech to the nation, he formed a special commission to investigate the cause or causes of the Challenger accident. He appointed former Secretary of State William Rogers to be head of the commission. The mandate of the commission was to review the circumstances surrounding the accident, to establish the probable cause or causes of the accident, and to develop recommendations for corrective action based upon the commission's findings and determinations.

Within minutes of the Challenger explosion, the Kennedy Space Center was in virtual lockdown. All personnel involved in the launch activities could not leave the center. All records and data was gathered and put under lock. At the other NASA Space Centers that had shuttle-related activities, action was taken almost immediately to form investigative groups relative to their activities. At MSFC, teams were quickly formed to investigate the shuttle elements they were responsible for. There was a team to review the external tank, the solid rocket boosters, and the main engines. Since I had been involved with many aspects of the external tank, I was assigned to that team chaired by Harold (Hal) Coldwater. There were two of us from the Material and Processes Laboratory (M&P) on the team. Dr. James Stuckey was the other M&P representative.

We worked twelve to sixteen hours a day reviewing in detail all of the manufacturing and launch data and launch films taken from different locations at Kennedy Space Center. Occasionally we would get calls from members of the Rogers Commission wanting information. Since I had alternated with Billy K. Davis on the launch Ice/Debris Team at one point during the review, I was interviewed by telephone by a member of the Rogers Commission. I was informed that I was being recorded during the interview.

On May 2, 1986, I received a letter, which stated,

> Shortly after the loss of the STS 51-L Mission, the External Tank Contingency Team was activated to determine the extent of the External Tank (LWT-19) contributing to the accident. Mountains of documentation, drawings, engineering changes, and material review actions were reviewed and reported on; many detailed analyses were performed and compared to tests and flight data; and key tests were run to provide substantiating data. All of this was done under very tight schedules to support key meetings with members of

> the Presidential Commission and to provide evidence of the External Tank not being the initiating element in the loss of STS 51-L.
>
> Your dedication and personal sacrifice as a member of the Team were prime contributors to the success of our investigation and in the development of the External Tank final report. I want to express my appreciation to you for your outstanding and unselfish support during that very trying time. I'm sure our efforts will be rewarded by successful recovery from this ordeal and we will have proud moments to look forward to in the future.

The letter was signed H.R. Coldwater, Chairman, External Tank Contingency Team, STS 51-L Accident Investigation Team.

Commission Findings

After exhaustive investigations, the Rogers Commission concluded that the cause of the Challenger accident was the failure of the pressure seal in the aft field joint of the right solid rocket motor (SRM). The failure was due to a faulty design and unacceptable sensitivity to a number of factors. These factors were the effects of temperature, physical dimensions, and character of materials, effects of reusability, processing, and the reaction of the joint to dynamic loading.

The commission determined that there were multiple contributing causes of the accident; what some call a *stacking effect*. They concluded that the decision to launch the Challenger was flawed and that the waiving of launch constraints appeared to have been at the expense of flight safety. They also found that the analysis of all the testimony and interviews established that Rockwell's (Orbiter contractor) recommendation on the launch was ambiguous and that the NASA Level I and II (Headquarters and Project office, respectively) response to the Rockwell position about the ice conditions was inadequate. The commission also found that the accident was rooted in history, beginning with decisions made in the design of the joint and in the failure by both Thiokol and NASA's SRB project office to understand and respond to facts obtained during testing. They determined that the joint test and certification program was inadequate, in that there was no requirement to configure the qualification test motor as it would be in flight. Neither NASA nor Thiokol fully understood the mechanism by which the joint sealing action took place. Even after several hot-gas, O-ring blow-bys (where hot burning gas escapes around the O-ring) in the joint during previous launches, NASA and Thiokol failed in their decision making by accepting the risk, because they got by with it before. The O-ring erosion and blow-by history had been reported to NASA Level I and II with no action taken. The commission determined that a careful analysis of the flight history of O-ring performance would have revealed the correlation of O-ring damage and low temperature. Neither NASA nor Thiokol carried out such an analysis to evaluate the risk of launching the Challenger in conditions more extreme that they had encountered before.

Further, the commission found that NASA had what they called a "Silent Safety Program." The Safety,

Quality Assurance, and Reliability organizations at both Marshall Space Flight Center and Kennedy Space Center had been placed under the supervision of the very organizations and activities whose efforts they were to check and oversee. That was like putting a fox in a chicken house.

During their investigation, the commission found that there was significant pressure on the whole Shuttle Program. Early plans projected one launch a week during the operational phase of the program. By 1985, NASA had revised the flight rate to twenty-four flights per year by 1990. Long before the Challenger accident, it had become obvious to NASA that two flights per month were overambitious. The reasons were largely based on spare parts availability, changes to major payloads, and changes to missions that adversely affected the astronaut training and development of procedures prior to the missions. In addition, the physical strain and mental fatigue on launch support personnel was a problem. Therefore, the previously scheduled launch rate did not accurately reflect the capabilities and resources.

Recommendations

Based on the commission findings and determinations, they recommended that the faulty field joint and seal must be changed. It could be a new design eliminating the joint or redesign of the current joint and seal.

The commission suggested that the National Research Council should form an independent solid rocket motor design oversight committee to implement the commission's recommendations. They further recommended that the whole Shuttle Program be reviewed and that the managers of the various shuttle elements be more accountable to their center directors than to the program organization. They also felt that it was important to use astronauts in essential management positions.

In addition, they felt that NASA should form a Space Shuttle System Safety Advisory Panel, which would report to the program manager. They also indicated that NASA should establish an Office of Safety, Reliability, and Quality Assurance, to be headed by an associate administrator, reporting directly to the NASA administrator; and that office should have independent oversight of all NASA functional and program responsibilities.

Other recommendations were directed toward better communication between NASA and contractor organizations bearing on safety of flights, to include establishing a system of analyzing and reporting performance trends; and establishing a flight rate compatible with resources and a firm payload assignment policy.

Follow-Up

Immediately after the Rogers Commission released their report, many things started to happen. As usually occurs after a national disaster of any kind, the news media and the government started looking for someone to blame.

At MSFC, Dr. William Lucus, the center director, and Lawrence (Larry) Malloy, the SRM Program lead, resigned, most likely under duress. J.R. Thompson was appointed in September 1986 to take the lead at MSFC as director.

At NASA Headquarters, George Rodney, a former Martin Marietta vice president, was appointed as the associate administrator for SRM&QA (Safety, Reliability, Maintainability and Quality Assurance). Jim Ehl, a MSFC division chief, was appointed as his deputy.

After the Rogers Commission made their recommendations, the NASA administrator appointed Admiral Richard Truly as associate administrator for Space Flight and Chairman of a Data and Design Analysis Task Force. Early on, Admiral Truly appointed J.R. Thompson as his vice chairman in charge of NASA's part of the accident investigation. J.R. had worked for NASA before but had gone to work in private industry for several years before being called back to NASA.

Within days after the commission report went public, J.R. Thompson and James (Jim) Odom, the MSFC Director of Science and Engineering, called Robert (Bob) Schwinghamer, the director of the M&P Laboratory. They requested that he appoint someone to chair a team to review the contamination and bonding process controls at Thiokol's SRM manufacturing facility in Utah. Since the buildup of the solid rocket motor is dependent on the bonding of the internal rubber lining and of the seals of the field joints, it was felt that it was critical that all of those processes be reviewed. The review was to be part of a redesign effort chaired by John Thomas and Royce Mitchell. John Thomas had replaced Jim Kingsbury, the previous deputy director of MSFC who was said to be, "too bullish on return to flight."

Bob Schwinghamer immediately called me to his office and told me to put together a team with the necessary expertise to perform the review, as soon as possible. Within three days, I had selected nine other MSFC employees that I knew had vast experience in adhesive bonding operations of all kinds.

Meanwhile, the SRM Project Office called the senior manager of Thiokol's Wasatch Division, relating to him that we were coming. We departed a week later, prepared to stay at least two weeks in Utah.

Upon our arrival at the NASA resident office at Thiokol's Wasatch facility in Utah, we were informed by John Thomas that we had better be prepared for a cool reception by Thiokol management. We asked the NASA resident office representative to have Thiokol management set up a meeting between our team and their manufacturing supervision, where we would relate to them the support we desired for the review.

The next day were taken to a conference room, complete with every manager and supervisor at the Thiokol facility, as well as John Thomas and Royce Mitchell. The reception was very cool, to say the least. It was obvious that they were really feeling the pressure of the news media, as well as NASA, because their hardware was found to be responsible for the death of seven astronauts. I felt sorry for them in a way.

The meeting started out by their senior manager saying that they had enough problems and didn't have the time to spend supporting our review. I informed him that we were there under direction of Admiral Truly's associate director and that if he couldn't support us, he needed to call J.R. Thompson. He then asked what we needed to get started, and we went from there.

Over the next two weeks, the team reviewed all of their bonding process documentation, including process procedures and controls. We then went to the manufacturing facility to observe just how closely the paperwork was being followed by the technicians.

Each morning during the review, the team started the day's activities out in a meeting, to coordinate what we were doing. It was agreed by the team that at the end of the review, I would present our findings to Thiokol's management before departing.

On the day we were to depart, we gathered in a conference room with only Thiokol's top managers. As I presented each of our findings and made recommendations for what the team considered viable solutions, their senior manager would dispute them. I had cautioned the team that we would probably get some resistance during the presentation and to not get into an argumentative mode. I also told them that when I finished, we would go immediately to our cars and leave for the airport. Just after I presented our last finding and recommendation, their senior manager was on his feet. We put our paperwork in our briefcases and started down the hallway toward the door. Their senior manager followed us down the hallway, still trying to argue the points of the presentation. We did not reply in any way to his tirade.

The other team members didn't know it, but almost every night during the review, I would talk with Jim Odom by phone about the way things were going with the team. Upon returning to MSFC, I schedule a meeting with J.R., Jim, and Bob to present our findings. The whole review team attended the meeting in J.R.'s conference room.

I made the same presentation I had made at Thiokol. J.R. thanked the team for their effort, and I was asked to stay a few minutes longer. As the other team members departed, J.R. said, "Charlie, this is what I want you to do. I want you to go back out to Thiokol tomorrow and snoop around and see what Thiokol is doing to implement your recommendations. Stay in touch, and Jim, Bob, and I will fly out in a few days for a conference with Thiokol's management."

I returned to Utah the following day. The next morning, I walked into the NASA resident office, and they were certainly surprised to see me. I was informed that Thiokol's senior manager was very upset with our team. I asked if Thiokol was taking any action to implement our recommendations and was told that there was nothing being done. Over the next few days, I attended some SRM Program update reviews with the NASA resident office personnel, only to receive some glaring looks from some of the Thiokol managers. Thankfully, their senior manager was not in any of the meetings. As directed, I would report to Jim and J.R. every night via phone. The last night, Jim said they would be in Utah the following day to meet for a conference.

Upon their arrival, I went to their motel and J.R. had arranged a conference with Thiokol management at their Management Training Center. He said that after I made the presentation that I had made before, to just sit down and not argue with the Thiokol managers, and he would take it from there.

The next morning as I was driving up the Ogden Canyon, I was thinking about just how beautiful the drive was. There was a stream running down the canyon and high cliffs on both sides with an occasional cabin alongside the stream. I passed a restaurant (Grey Cliff), where I had eaten before, and was thinking about how good their trout specialty was as I drove by.

The Thiokol Management Training Center was initially an old rustic lodge and as I approached it, I was thinking about how well built and beautiful it was. Even with all the beauty of the canyon and the lodge, I knew that things were probably going to get ugly inside. I expected that the company senior manager would interrupt my presentation, as he had done the first time.

I was pleasantly surprised when the meeting got underway; the Thiokol managers were all very polite and cordial. As I made my presentation, no one interrupted or disputed anything I said. After I finished, I gathered my viewgraphs, put them in my briefcase, and sat down.

Almost immediately, J.R. turned his attention to the senior manager, who sat there with a slightly red face, and asked, "What are you doing to implement Charlie's recommendations?"

With his face getting even redder, he replied, "We have been building rockets the way we do for over thirty years, and we don't intend on changing anything now!"

J.R. responded, "I think you should reconsider that position!"

The manager's only response was to say, "Are we through?"

J.R. replied, "I guess so!"

The meeting was adjourned and as we were departing, J.R. mentioned that I could ride back to MSFC on the NASA airplane. We drove to the local airport, boarded, and flew directly back to MSFC. Nothing was said of the meeting with Thiokol on the flight back, but I could tell that J.R., Jim, and Bob were all in deep thought.

About a week later, Bob called me and asked that I accompany him to J.R.'s office. When we got there, Jim Odom was already there.

As we entered the room, J.R. said, "Come on in! We have some serious days ahead of us."

When he had finished looking at some papers in front of him, he looked at me and said, "Charlie, I want you to divide your team into two groups and go back to Thiokol and work with them in implementing your recommendations. You go out with the first group and coordinate the activities with Thiokol management. Rotate the groups every two weeks until your recommendations are implemented. If you need anything, just call Jim or me and you will get what you need."

I asked him how quick we should go to Thiokol and he told us as soon as possible. J.R. then turned to Jim Odom and started talking about something.

As Bob and I left, I started dreading the return to the Thiokol facility. I knew that I had a major task ahead of me considering the adversity of their management.

Jim caught up with us before we got to the elevator and brought us up-to-date on what had transpired after the meeting at Thiokol's Management Training Center a week earlier. J.R. had called the Morton Thiokol corporate office upon returning to MSFC and talked with the chairman of the board, relating to him the problem he had with their top manager at the Wasatch Division in Utah. Jim told us that the senior manager had been replaced with another from the home office in Connecticut and that was who I would be working with to implement my set of recommendations. That news certainly relieved my mind considerably.

Two days later, I took the first group back to Utah and met the new manager that had replaced the previous. He formed a team of lower-level managers to work with us and we got started. We knew that it would take at least two years to implement all of the recommendations.

Redesigned SRM

While John Thomas and Royce Mitchell were working with Thiokol to redesign the SRM clevis joint and seal, which caused the Challenger accident, another major project was started with headquarters' blessings. It was a totally redesigned solid rocket motor (RSRM).

MSFC's engineers went to work on the design of the new RSRM system and MSFC management acquired an abandoned Tennessee Valley Authority (TVA) nuclear power plant site at Yellow Creek, near Iuka, Mississippi. The site was to be utilized as the manufacturing facility for the RSRM by a different contractor other than Thiokol. The RSRM would not have as many field joints and would be tested in the vertical position, unlike the Thiokol-built SRM. Jim Blair, the director of the MSFC Structures and Dynamics Laboratory, directed the overall effort.

Since I was heavily involved with the production processes that Thiokol used, I was asked to participate in all of the RSRM reviews as the design progressed. I was also on the team that reviewed the TVA site to determine if any of the structures could be used in the RSRM program.

There was more than one objective of the RSRM Program. The first was to meet the recommendation of the Rogers Commission and second, to keep pressure on Thiokol in the short haul to clean up their processing and to keep the space shuttle flying until the RSRM came online. As a matter of fact, on two separate occasions, the manager that had replaced the top manager of the Thiokol Wasatch Division asked me, "What will it take to shut the Yellow Creek RSRM Program down?"

Each time I told him that Thiokol would have to show hard proof to NASA management that they had the problems corrected that caused the Challenger accident.

I don't know how much pressure the Utah delegation applied in Washington but eventually, the RSRM program at the Yellow Creek site was stopped.

During the recovery period after the Challenger accident, Dr. James Fletcher was the NASA Administrator (May 12, 1986–April 8, 1989). It's a fact that during that time, he took a lot of heat from the news media because he was of the Mormon faith and from Utah. With the space shuttle SRM contractor, Thiokol, being located in Utah and a large number of the employees being Mormons, things did look suspicious.

TVA spent around two billion dollars at the Yellow Creek site before they discontinued work there. NASA spent in excess of another billion at the site before it was abandoned again. It's a shame that all that money was spent, but I suppose that is not the only funds wasted by our government.

After the RSRM project was stopped, I received a letter that stated,

> I would like to express my great appreciation for the excellent work you did on the RSRM Case-Nozzle Joint Assessment Team. Successful completion of our task required an extensive effort, digestion of a large amount of information in a short time, and significant judgment in forming conclusions.
>
> As Materials and Processes representative, you brought a broad and detailed experience base to bear on the issues. Your observations, insight, and judgment were essential elements of the team activity. Your contributions are greatly appreciated. Many thanks for a job well done.

The letter was signed by James C. Blair, director, Structures and Dynamics Laboratory.

An amusing thing happened at MSFC during the activity to return the space shuttle to flight status. In an effort to raise morale, J.R. Thompson wanted to have a really good Christmas reception for the MSFC employees. Jim Odom called me and said that J.R. wanted to know if I could have Martin Marietta, the contractor that operated the MSFC Spray-On Foam Insulation (SOFI) facility, build a foam snowman for the reception. I assured him that I would take care of the matter. I called one of my branch employees, Jim Thaxton, and related to him what I wanted, and for him to work with Martin Marietta to get the job done. As it turned out, they made three snowmen instead of one, and all three were huge. J.R. Thompson and Jim Odom were astounded when they saw what had been done.

I don't know who had it done but after the reception, the three snowmen were given a bar code capital equipment number and put in a capital equipment warehouse. One of my contacts at MSFC told me that they had been used at each Christmas reception for many years after my retirement. When Jim Thaxton retired years later, the Martin Marietta people presented him with a miniature foam snowman in remembrance of the event.

Snow Men

Called to Headquarters

I had been rotating back and forth to Utah with my group for about ten months when my life changed. By November, Thiokol had implemented several of the recommendations and everything was going fine. During a meeting with Thiokol personnel, I was interrupted to answer a phone call from NASA Headquarters. The call was from James (Jim) Ehl, my former MSFC division chief, who was now the deputy to George Rodney, the new associate administrator for SRM&QA, for the Agency. Jim said that he and George were looking for someone to fill the position of director of SRM&QA for the International Space Station program, and my name kept coming up. He said that I had about a week to decide if I wanted the job, and that the job would be an Excepted Service position with pay equivalent to that of a GS-15. He said the job was to represent the Agency as the Headquarters lead for the Space Station SRM&QA activities with the NASA Field Centers, international partners, Department of Defense, and other government agencies, as well as oversight of Space Station contractor organizations. He also said I would be the principal in defining organizational roles and responsibilities for Space Station SRM&QA organizations with NASA.

Needless to say, I was totally surprised and dumbfounded by the offer. I immediately called my wife and told her of the offer and that we would talk some more when I returned home that weekend. After returning home we discussed the offer and decided that if I didn't take the offer, in the future I might look back and wonder, *What if?* The following Monday I called Jim and told him I would accept the offer. He said that he would get things going at headquarters and I could do the same at MSFC. I informed my management and things happened quickly.

In January of 1988, I reported to George Rodney at NASA Headquarters and was assigned an office at the Space Station Project in Reston, Virginia, and in the Headquarters building across the street from the National Air and Space Museum in Washington, D.C. I had two offices, because I was responsible for both Level I and II Space Station SRM&QA requirements. I would spend time in both offices, as needed to do the job. The first two days in the new job was utilized mostly settling into an apartment, finding my way around, and meeting the NASA and contractor people that I would be working with.

The third morning on the job, I got a call from Jim Ehl just after I arrived at my office in Reston. He said that I needed to come to the Headquarters building immediately. I went to the Metro station in Fairfax and less than an hour later had arrived at the L' Enfant Plaza Metro Station across the street from the NASA Headquarters building.

Upon reaching Jim's office, his secretary told me to go on in. Jim told me that George Rodney was scheduled to meet the astronauts at the Johnson Space Center the next day, but that he was sick and could not go. He said that he could not fill in for George because his father in Gadsden, Alabama, was ill and he had to go there. He said that he and George had decided that I was the one to go because of my involvement in the Thiokol SRM processing cleanup and the RSRM Project. He said the astronauts wanted an update of the progress at Thiokol and to discuss the return to flight. George's secretary already had travel orders for me to fly early the next morning. I returned to my office in Reston and spent most of the day preparing viewgraphs for my presentation to the astronauts.

Early the next morning, I flew to Houston and drove to Johnson Space Flight Center. I went directly to the astronaut office and was escorted to a conference room where the full Astronaut Corp. was waiting. I walked into the conference room, introduced myself, and then proceeded to explain to the Corp. why I was there instead of George Rodney or Jim. I told them that I had been selected to meet with them because of my involvement with Thiokol's cleaning and bonding operations on the SRM and with John Thomas's SRM redesign team at Thiokol. Richard Covey immediately told me that they knew who I was and about the MSFC team that I had taken to Thiokol and the follow-up team work with Thiokol. He then introduced the other astronauts and told me to proceed.

I made my presentation as I had prepared, occasionally stopping to answer questions. At the very end of my presentation, Fredrick (Rick) Hauck said, "What we want to know is, are we going to be safe to fly again?"

After thinking for a moment, I responded, "I feel that what John Thomas and Royce Mitchell are doing with Thiokol on upgrading the SRM clevis joint O-rings and seals and with what the Contamination and Bonding Process Review Team are doing, the SRM will be as safe as we can make it. Even with all that effort, there will always be some risk remaining."

Rick Hauck responded, "We all understand that!" After that exchange, he thanked me for coming and I started breathing again as I departed for Washington.

Back in Washington I started looking for personnel to fill the positions in my organizations and making contacts with the SRM&QA organizations at all the Field Centers. Within days, I had a secretary and managers for the Safety, Quality Assurance, and Reliability/Maintainability offices. And after few months,

I had filled the other positions authorized for my Space Station Safety and Product Assurance Office. I had hired people from private industry and internally from NASA offices. Most had many years experience in their fields.

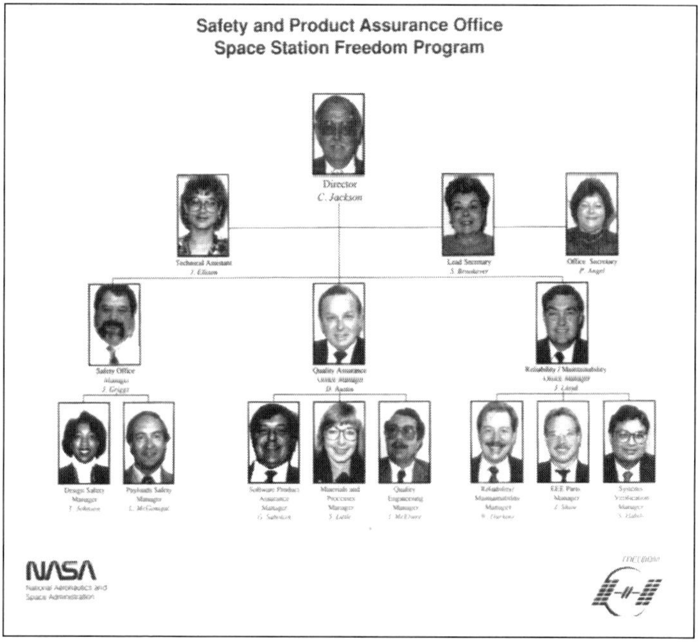

Space Station SRM&QA Organization Chart

Shortly after getting started in the new job, I called a meeting with my counterparts from the Space Station international partners and with the SRM&QA directors of the NASA Field Centers. The meeting was primarily for getting acquainted and for establishing guidelines for how we were going to work together.

During that first meeting with the international partners, I was amazed at how my counterparts looked at me. Mr. Gono, from Japan, was a very humble and polite person, while Dr. Hendrick Wessels (pronounced *vessels*), from the European Space Agency, was very arrogant. He let me know right off that he was to be referred to as "Dr. Wessels."

During the meeting, the representatives from the NASA centers started arguing about who was going to be responsible for various things. Johnson Space Center thought that they should be the lead center, since they were the manned flight center for NASA's previous flight programs. They really didn't see any need for my Level II organization at Reston. The arguing was about to get out of hand when I asked the international partners to step out of the conference room.

Closing the door, I informed the representatives from the NASA Centers that we were not going to argue about our internal problems in front of the international partners. I told them that if that was all they could do, they could pack their bags and go home. I told them the decision to have the lead in Reston was made at the NASA administrator level and if they disagreed, they should take it up with their higher management.

"Now, can we work together?" I asked.

After a moment of deathly silence, the Johnson Space Center (JSC) representative responded with an okay. I called the international partner representatives back into the room and from there on, we had a productive meeting. I realized much later in the Space Station program that the attitude of the JSC representatives at that meeting prevailed pretty much throughout the JSC management.

Not too long after I left MSFC, J.R. Thompson was reassigned to NASA Headquarters as deputy associate administrator to Admiral Richard Truly, who had been appointed by the President of the United States as NASA Administrator to replace Dr. James Fletcher, as recommended by the Rogers Commission. Jack Lee was appointed director of MSFC when J.R. left for Headquarters. Jim Kingsbury, the deputy director of MSFC, had been replaced by John Thomas. After Jack was appointed director of MSFC, he called me and said that he would leave a light burning in the window for me if I wanted to return some day.

Several months later, I received another phone call from Jack Lee. He said that I should return to MSFC on a certain day to attend an awards ceremony but would not tell me why.

I flew to Huntsville, Alabama, the day before the ceremony was to take place. During the ceremony I was asked to come forward and when I did, I was presented a second NASA Metal for Exceptional Engineering Achievement. The award certificate stated that it was presented for exceptional engineering achievement in establishing an in-depth contamination control protocol for solid rocket motor fabrication.

During my tenure at Reston and Washington, I participated in numerous activities. I attended the

Space Station signing ceremony, by the international partners and NASA, at the state department with then Secretary of State George Schultz, and I met several of the representatives of the international partner nations.

I participated in numerous Space Station Progress Reviews with some congressional staffers in attendance. One that I remember in particular was where a congressional staffer for Senator Mikulski from Maryland was in attendance. Richard Tinius, previously of MSFC, was presenting the assembly sequence for the Space Station when he was interrupted by the staffer who said, "The senator wants the piece of Space Station hardware that Goddard Space Center (Goddard is in Maryland) is responsible for to be the first part of the Space Station put into orbit."

Richard responded by saying, "That would be nice, but we can't do that."

The staffer responded, "But you don't understand! This is Washington and what the senator expects, she gets."

Then Richard replied, "You don't understand! If we launch that piece of hardware first, it will start tumbling and fall back into the earth's atmosphere and burn up because it does not have a gyroscope aboard to keep it from starting to tumble."

During my tenure in Washington, I found that many of the congressional staffers are young lawyers that have little technical knowledge. What is really scary is that they are the people providing information to the congressmen and senators on which decisions are made that impact the whole nation.

I attended the parties given by the international partners when they opened their respective resident offices in Reston, Virginia. They certainly didn't scrimp on the budget for food. The Japan party was probably the best, in my estimation. I always had to be careful when attending parties, meetings, and reviews with contractors and international partners, that I did not get compromised by accepting meals and gifts which exceeded what was allowed by government regulations.

I attended budgetary hearings with George Rodney at the Rayburn Building with the Congressional Budget Subcommittee and got to meet several congressmen and many of their staffers.

One day when I had time, I went to Alabama Senator Howell Heflin's office to meet him. I told him what I was doing in Washington and when I told him that I was from his hometown, he really relaxed and wanted to chat. When I told him that his wife (before he married her) was one of my teachers at Deshler High School, he got excited. He was really taken aback when I told him that my wife's aunt was his office manager in Tuscumbia. We chatted for some time before he had to leave for a meeting.

Within the first year that I was in Reston and Washington, Jim Odom was transferred from MSFC to NASA Headquarters as the associate administrator for the Space Station. Since I had known Jim for many years through our church in Decatur, Alabama,

I had a good friend that I could consult and socialize with from time to time, which made my time there much more pleasant. I had dinner with Jim and his wife, June, a number of times at their apartment.

During my time in Washington, I traveled to Toronto, Canada, and Nordwick in the Netherlands where the European Technical Center was located, and to Tokyo, Japan to negotiate the Space Station SRM&QA requirements for the program. I visited most of the NASA Field Centers that had a part in the Space Station to coordinate their involvement in the program activities and to assure that they understood the requirements.

My wife, Judi, and I enjoyed the many cultural activities that northern Virginia and Washington, DC, offered. We spent almost every weekend visiting some historical site in and around the area. We visited Mt. Vernon, all the Civil War Battle Fields, including Gettysburg, Pennsylvania, Harpers Ferry, and most of the many national museums. We attended many concerts in Wolf Trap and around the Washington Monument and on the west lawn of the Capitol Building. We attended the Fourth of July celebration on the west lawn of the Capitol and the National Mall. There was almost always something going on around the Mall.

The apartment where we resided while working in Washington was located in Oakton, a small community between Vienna and Fairfax, Virginia. We had a famous and interesting neighbor living in the same apartment complex. He was Fred Haise, the astronaut of the

Apollo 13 near tragedy. Fred was one of only twenty-four astronauts to go to the moon but was unable to land because of an oxygen tank explosion shortly after launch. He and his companions endured three days of harrowing survival during their return to earth.

When I went to work in Reston and Washington, Fred had quit NASA and was working for Grumman Space Company as a contractor on the Space Station program, also in Reston. On occasions my wife and I would see Fred when we were on nightly walks around the complex. My wife and I always enjoyed talking to Fred about his experiences. He had survived Apollo 13, an airplane crash, and a serious automobile wreck. In the Space Station Project Office in Reston, he was often kidded about already using three of his thirteen lives, like he was a cat.

Since I had an office just across the street from the National Air and Space Museum, I often ate lunch in the cafeteria of the museum and stayed up-to-date on the exhibits there.

During the two years that I was in Reston, Virginia and Washington, DC, there was involvement (a better term would be meddling) from Capitol Hill congressmen and their staffers in the Space Station program. At the same time, Johnson Space Center management was constantly trying to get the program moved to their center.

Return to MSFC

In April of 1990, I returned to Marshall Space Flight Center as Deputy Director of the Non-Metallic Materials Division of the Materials and Processes Laboratory. Jim Lloyd, the Reliability and Maintainability Manager for Space Station, took my previous position as director of SRM&QA for Space Station.

About a year and a half later, the management of the Space Station program was moved to Johnson Space Center from Reston, Virginia. It was probably the best thing that could have happened, because it removed the program from the constant scrutiny and meddling of the Washington politicians. In the move, the Reston, Virginia Space Station program organizations were broken up and the personnel reassigned to other NASA Field Centers or Headquarters organizations.

On April 18, 1990, I received a letter which stated,

> In accordance with your request to return to Marshall Space Flight Center as Deputy Division Director, Non-Metallic Materials Division in the Materials and Processes Laboratory, your Excepted Appointment will be terminated effective April 22, 1990. During the period of your appointment you have made a significant contribution to the Space Station Freedom Program and to this office as well. I appreciate the tremendous effort you have put forth and wish you well in your new endeavor.

The letter was signed, George A. Rodney, associate administrator for Safety, Reliability, Maintainability, and Quality Assurance.

Ron McIntosh was director of the division in which I was assigned at MSFC. Ron was a laid-back person and didn't let much of anything bother him. He liked to joke around with some people from time to time and would get a kick out of a good practical joke, as long as the job got done.

On one occasion, I had gone to Kennedy Space Center to represent the division in a project review, and Ron called to say that he needed me to send him a status report of what was happening at the review. I told him that I hadn't done anything since I had been there and the telephone line instantly got silent.

After a long pause, he said, "Well! I really need something."

He didn't know it, but I had already faxed a report to his secretary. After he hung up, I called his secretary and asked her to take a copy of the report to his office and lay it on his desk.

In about fifteen minutes, he called me back and said, "You son-of-a-gun, I believed you. I was really sweating what I was going to put in my report to the boss."

I found that was only one of many jokes between us while I worked for Ron. Most of the time I was on the receiving end of the jokes.

While serving as the deputy of the division, I had many duties. I wrote a lot of annual evaluation reports for division employees, often including my own. I attended numerous meetings for Ron when he had time conflicts. Since I had been involved with the MSFC PEC from the beginning, I was the primary division interface with most of the contractors in that facility. I addition, I was involved in the periodic contractor evaluations for that facility.

One of the most time consuming jobs I had was chairing accident investigation teams for MSFC. I seemed that every time there was need for an investigation, I was assigned the job of chairing a team. I served as chairman of three teams in one year. In the absence of a branch chief, I would fill in as the supervisor. Some of the periods were fairly lengthy, such as when a branch chief was on vacation or on a long-term assignment.

In 1994, President Clinton announced that his administration was going to reinvent government. As part of that program, there was a concerted effort within NASA to move more women and minorities into upper-management positions. Almost all of the men that were being "groomed" for higher management positions were sidelined as a result of this effort.

NASA announced a buyout program for those employees with numerous years of service for several reasons. They would try to get rid of high-salaried employees so that they could hire lower-salaried personnel in their place. Also that would get the older employees out of the existing civil service retirement system (CSRS) and put younger employees in a retirement system, consisting primarily of Social Security, with a minimal Federal Employee Retirement System (FERS).

Early in 1994, MSFC announce a center-wide reorganization that would do away with most division deputies. Seeing the writing on the wall, I took the NASA buyout and retired on May 4, 1994.

Consulting

After six months of doing almost nothing after retiring, I decided, with urging from NASA friends, that I would get into the aerospace consulting business. The only thing that my wife and I were doing that might interfere with working in the consulting field was some serious RV camping. I had bought a fifth wheel and a pickup truck in 1992, and we were spending lots of time on the road and at the Gulf of Mexico.

In late fall of 1994, I formed a subchapter S corporation, C.J. Associates, Inc., in the state of Alabama and was awarded a contract with NASA to provide assistance in a technology transfer effort. McDonnell Douglas in Pueblo, Colorado, wanted to use a spray-on ablator, developed by MSFC for the space shuttle SRBs, on their Delta II rocket system. During the contract period, which lasted eighteen months, I worked as average of two days per week at MSFC and spent another day per week at home writing reports, etc., and getting them professionally prepared for presentation to the NASA contracting officer. Occasionally, I would work a full week when required to travel to Colorado for an implementation status reviews.

Near the end of that contract, I won another contract with United Space Boosters Incorporated (USBI) for technology transfer of the Ablator system to Boeing (later to become United Launch Alliance) in Decatur, Alabama. The work was similar to the previous NASA contract and lasted another eighteen months. After the completion of that contract, we caught the travel bug and I decided to quit work altogether and devote more time to RV travel.

Travel

During summer of 1998, Judi and I bought a thirty-two foot motor home and a small car to pull behind it and hit the road. We helped form a travel club and started camping every month of the year, all around the southeastern part of the United States. The manufacturer of the motor homes would sponsor state and national rallies, many of which we attended. One time while we were with our club, a man asked us what we did as a club. Without giving it much thought, I replied, "We just travel around, hunting nothing to do." He laughed and said that he could get into that mode easily.

For the next two years, we camped every month with the club and alone as much as possible. We spent many days on the Gulf Coast soaking up the sunshine, saltwater, and eating a lot of seafood.

36 *Foot Motorcoach*

Thirty-Six Foot Motorhome

In early 2000, we traded the thirty-two-foot motor home for a thirty-six-foot motor coach, with a slide-out, which provided much more internal space. We traded the small car for a small pickup to pull behind the larger motor home. Since the thirty-six-foot motor coach was more powerful, it was easier to drive on longer trips than before. We made trips to Michigan, Missouri, Texas, Illinois, and Nova Scotia in Canada.

The Nova Scotia trip in 2003 lasted six weeks and one day. We went into every New England state except Delaware. The trip began on the eighth of September, and we returned home at the end of October, following the fall color change back south.

Two other couples, James and Laverne Roberts of Decatur and Joe and Caroline McCullough of Clarksville, Tennessee, made the Nova Scotia trip with us. We caravanned with our RVs, alternating the lead couple each day. We all had CB radios and if the lead couple got lost, the other couples would start singing the *Wild Goose* song over the CBs.

A few times, we took those big rigs through the cobblestone streets in New England, which were not meant to accommodate any vehicle larger than a wagon. The rearview mirrors on the sides of the motor coach would nearly clip power poles on either side of the narrow streets. We would get some strange stares from the locals.

We had one incident with a trucker while driving back to the US through New Brunswick. We were caravanning, cruising along at about sixty-five miles per hour when I noticed that a trucker was getting agitated and trying to pass. I was in the second position when the trucker passed the caravan. He immediately whipped over in front of the McCullough's, causing them to swerve and almost lose control of their RV. Just as he dropped into the front of the McCullough's, he hit his brakes, which could have potentially caused a wreck. I got on my CB radio and told Joe to switch to the channel most used by truckers. We switched and started talking about the trucker, knowing that he could probably hear us. The trucker would hit his brakes, still giving us a hard time.

I said, "Joe, get his tag number, and we will report him to the authorities to get his license revoked!"

I had no more than finished saying that when he floored that truck and with black diesel smoke curling in the air, he was quickly out of sight.

I noticed during our RV years that most truckers were not enamored with people in RVs of any size. I never did understand why, unless it is because some travel as a caravan.

Personal Tragedy

Beginning around 2002, I noticed some swelling in my legs on bus tours. I just thought it was caused by sitting so long and the pooling of fluid in my legs. In 2003 I noticed pronounced weakness in both legs, especially going down steps or slopes. I had to be extra careful or my legs would collapse. During the RV trip to Nova Scotia, I fell to my knees when descending some stairs in the Citadel of Halifax. Fortunately, I was holding to the handrail. Later the same day, we were going downhill, and I had to hold onto my wife's shoulder to keep from going down again. I knew I had a serious problem.

When we got back to Alabama, I made an appointment with my family doctor. After a thorough physical, my doctor referred me to a neurologist, thinking that I might have Multiple Sclerosis or even Lou Gehrig's disease. After being examined the neurologist made me an appointment with a muscle and nerve specialist at the University of Alabama Birmingham Hospital for a biopsy of nerve and muscle tissue. The specialist removed muscle and nerve tissue from my left leg and sent me home. After a week, I received a call from the neurologist asking me to return to his office. I was informed that the biopsies showed that I had a rare disease called polymyocitis, an immune system disorder in which the body's immune system attacks the muscles of the extremities and causes them to atrophy. The neurologist said that there was no known cure because it was so rare (only one in one hundred thousand), but the disease might be put in remission by treatment using massive doses of steroids. He immediately put me on a schedule of steroid treatment, starting with a dosage that I thought would burn me up from the inside. I now know how women feel when they have a hot flash. Over a period of several months, I completed the treatment but could not tell any difference in the strength of my legs.

In April 2004, Judi and I traveled by bus to Charleston, South Carolina, and Savannah, Georgia, with the seniors group from our church. We were traveling almost a week, and there was a noticeable reduction in swelling in my legs from previous bus trips. I attributed that to success of the steroid treatment.

Upon our return we were scheduled to attend a state level RV rally in Gadsden, Alabama. Our travel club was responsible for hosting the event for other camping club members from across the US and Canada. Friday, April 30, was the first official day of the rally, even though many RV rigs had been at the campground for several days. The attendees usually arrived early to catch up on news from other clubs and renew friendships with couples that they hadn't seen for years.

On Saturday morning, our club was set to start the rally games. Around nine a.m. the games were began and progressed smoothly on into the morning. At ten a.m. I told James Roberts it was time to pick up the meat for the evening meal. We left the campground in my truck and proceeded west toward Hartselle, Alabama, where the barbeque restaurant is located. When we reached the intersection of highways 278 and 77, we turned west on 278.

About a mile from the intersection, we were cruising along at fifty-five miles per hour when we approached a man in an eastbound van that had stopped, waiting for us to pass by, so that he could turn left. Just as we approached the van, a pickup truck, also eastbound, ran up behind the van at a high speed and tried to pass on the shoulder of the road. The truck clipped the right back corner of the van, causing it to spin ninety degrees and directly into our path, no more than fifteen feet in front of us. Without any time to react, we slammed into the van at fifty-five miles per hour. With a loud *bang*, the grinding impact was horrendous. I remember hearing the airbags pop open, glass breaking, and metal crunching.

I must have closed my eyes, because when the truck stopped moving, I saw that we were at least fifty feet back from the van. All of the glass was broken out of the truck, and the motor was driven back against the firewall. The steering wheel had me pinned against the seat back and center console.

There was lots of smoke rising just in front of me, where the windshield was before. I was sure that the truck was going to flame up at any moment. The thought of burning alive scared the daylights out of me. I tried to reach up with my right hand to turn the ignition switch off but found that my fingers would not work. I looked at my hand to see what was wrong and could see that my right wrist was offset and was beginning to swell grossly; I knew then it was broken. I then tried to reach around the steering wheel with my left hand. When I tried to lift it, the lower part of the arm, from the elbow down, would just flop down. I saw the elbow area was beginning to swell and knew that it too was broken.

The passenger side door was open, and James was able to get out of the truck and go to a nearby house to call for help. When he came back, he tried to help me get out but to no avail. I asked him to get my right shoe that was lying on the floorboard and keep it; why, I'll never know. James realized, somewhat later, that his left ankle was hurting him and realized that it was broken.

I hurt in both hips like I had never before hurt in my life. It was a deep burning hurt like someone driving nails into your bones or being cut through from hip to hip by a chainsaw. I also had a different kind of pain shooting down both legs to the tips of my toes. It was a burning sensation, much like an electrical shock, but more severe. I tried, but there was no way that I could move my legs. Strangely enough, my broken arms did not hurt at first but later joined the other parts of my body in sending pain signals to my brain.

Traffic stopped immediately. A man came up to where the truck window had been and asked me if I was injured. I told him I was and that I was pinned in and was afraid the truck was going to burst into flame. He assured me that the smoke was coming from the battery that had burst. He left and came back with a blanket or something, put it over the battery, and it stopped smoking.

Shortly after that, a fire truck arrived, and two firemen came up to the side of the truck. They asked me if I was injured, and I told them that my arms were broken and that I hurt terribly in my hips and legs. One of them put his hand on my left shoulder and asked if I could move my head. I moved it from side to side. Then I begged them to get me out of the truck, thinking it was still a fire hazard. They said they had to wait for the paramedics to arrive and check me before they could cut me out and move me.

I sat there and sat there for what seemed like an eternity. I had already asked my maker to just take me on and was becoming quite lethargic when the paramedics finally arrived. I found out much later that it was a forty-five minute wait and that I was bleeding profusely through my left elbow, because the bone was sticking out. The paramedics began checking me, and the firemen told me that they were going to start cutting me out of the truck, and it would be really noisy, like a jackhammer on metal. That is the last thing that I remember.

What I am going to relate to you now is second-hand information, because I do not remember anything for the better part of a month.

I was told that they took me to the hospital on a hill overlooking the Coosa River in Gadsden, Alabama. Apparently, I talked to the paramedics, doctors, my wife, and some camping friends but don't remember any of it. I had read somewhere that the human body could only withstand so much pain before the brain would shutdown, no longer recognizing it. I firmly believe that now.

Judi said that I was flown from Gadsden, Alabama to Huntsville, Alabama, by medical helicopter. She said that I made the choice between Huntsville Hospital and University Hospital in Birmingham, Alabama, even though I don't remember. According to her, I took an impact on the left side of my face, as it was black and blue. On the outside edge of my left armpit, I had a deep burn, caused either from the airbag or battery acid. My right wrist was broken, and the bones in my left elbow were broken and sticking out. My ribs on the left side were also broken, making it difficult to breath. The large bones on both sides of my pelvic area were broken vertically, and even overlapped on the right side. The sciatic nerve in the right pelvic area was crushed, leaving my leg paralyzed from the knee down. One of the eight doctors told her that I had one artery slightly nicked, which they repaired, but no other internal injuries.

Two surgeries, lasting sixteen hours, put my pelvic bones back together using twenty-two titanium screws. To make the needed repairs, the doctors cut me horizontally across the front of my body in the pelvic area and then at an angle up to the top of the left buttock. On the right, they cut me vertically from about halfway down the outer thigh to the top of the pelvic and then back on the buttock. I had over seventy staples in the surgical areas.

My left arm was also put back together using titanium screws and a metal plate. My left elbow had ten staples in it. The right wrist was reset, and a keeper bar was installed by putting screws into my hand and arm bones. The doctors were afraid that I would have blood clots so they put a permanent filter into the large blood vessel that returns blood to the heart and lungs from the legs.

I was in the intensive care unit (ICU) on life support for sixteen days, because I needed breathing support. For several days, they were not sure that I was going to survive. I remained in ICU until the twenty-second day, after being taken off of life support after the sixteenth day, and was then moved into a private room. I barely remember being moved. I was still heavily sedated and don't remember clearly until the first week at the nursing home.

While in the hospital, I only remember how grotesquely swollen I was and finally trying to get off the morphine drip. The second day after I came off of the drip was a Sunday, and I was told that I had a lot of company. I was a bundle of raw nerves, and my insides were shaking like Jell-O, and all I felt like doing was crying and screaming. I certainly was not in any shape for company.

Meanwhile, Judi and I had been building a new home and were scheduled to move on May 24. On the twenty-third, Judi was informed that I would be moved from Huntsville Hospital to a rehabilitative nursing home on the twenty-fourth. I think she went into the attack mode, because she had the doctors intercede with the hospital administration, and I was not moved out until after she had made the move into our new home. She said she had a lot of help from members of our church, both in making the move and sitting with me during the time she was taking care of things at home.

Recovery

On the twenty-sixth of May, I was moved into a room at Whitesburg Gardens, a nursing home in Huntsville, Alabama. I was put in the temporary wing of the nursing home, which meant you got better and went home, or you died while there. Actually, there were six deaths while I was there. About half of the occupants of that wing were terminal, and the other half were injured and were there to recover and take therapy.

My orthopedic surgeon ordered that I was not to put any weight on my pelvic areas for twelve weeks. That would give the broken bones time to heal before any stress was put on them. My wife said that twice while I was still in the hospital, therapists came into my room to get me up on my feet, and if she hadn't been there to tell them what the doctor had ordered, they would have.

My stay in the nursing home was an experience, to say the least. Since both my arms were broken, I couldn't do anything for myself. All I could do was lie flat on my back for the duration of the twelve weeks. Every morning I was awakened around six thirty by a night nurse taking my temperature and blood pressure. Then at eight, the day nurse would come in to administer my six pills and two shots into my abdomen. At eight thirty, the nurse's aides would bring me my stone cold breakfast, which I hardly ever ate completely. Shortly after breakfast, the nurse's aides would appear again to give me a sponge bath and change the sheets on my bed. I was usually stripped down to total nakedness and would lay there, completely helpless, before God and anyone—usually several aides and nurses who chose to walk into my room.

One day this male nurse's aide was giving me a bath, and he had raised my bed to the maximum height because he was a rather tall person. Judi walked into the room about that time and could not even see me I was so far up in the air. She said, "Are you up there?" and I replied, "I think so."

Lunch and dinner were experiences to behold. I never knew what I was going to get, and sometimes I didn't know what I had when I got it. One time I got what the nurse said was liver. It looked more like a shoe sole. I couldn't cut it or even bite a piece off of it. I just set it aside and Judi went to a local fast food restaurant and got me some fried shrimp. The nurses even complained to the kitchen personnel about them serving shoe soles for dinner.

After I had been in the nursing home for a few weeks, an area on my left hip began to bleed internally. It was very painful, and I was returned to Huntsville Hospital for nine days. After recovering, I returned to the nursing home to continue recovery.

When you are as severely injured as I was, it is physically impossible to go to a bathroom or even use a bedpan, because of the pain and the inability to use your arms. That is when you lose any modesty you ever had.

It seemed that June and July would never pass. Each hour seemed like a day, and each day seemed like a week. During the times that I was awake, I counted the ceiling tiles hundreds of times. Sometimes I would lay there listening to the other patient in the room snore or to people talking in the hallway. I often lay there straining to hear what people in the hallway were talking about. Some days I would have lots of visitors, and Judi was there almost every day to see that I was treated properly.

One day a lady, who was a patient, was standing outside my door ranting that God had spoken to her during the night and that someone in that wing of the nursing home was going to die that day. Now, that surely doesn't make one feel good. Another day that same lady came into my room, thinking it was her room. The nurse told her that it was not, but she did not want to leave. The nurse told her that it was a man's room and after she looked at me really hard, she said that she did not see any man there. At that point I thought she was going to get into my bed. The nurse had to take her by the arm and forcibly remove her.

I slept a lot during the day because I would be awakened so many times during the nights by nurses, aides, and by the other patient in the room. In addition, I was given muscle relaxants that made me sleepy.

Just lying there with nothing to do except think, I experienced just about every human emotion imaginable. I experienced anger, fear, anxiety, depression, despair, happiness, and a few others without names. Some days I was extremely low, hurting terribly and thinking that I would never get out of there. Those days of despair would turn into a desire to just pass on from this life. Often, just a slight shift of my body weight would temporarily relieve the pain in my pelvic area. During those times, my wife would pull up on the side of the bed sheet just enough to shift my body weigh in the opposite direction.

Other days I would feel better, not hurting as badly, and things would look brighter. It seemed that the longer I hurt, the more accustomed I became to the pain. Even that early after surgery, I could tell when a weather change was coming because of the increase in pain intensity. I still have that indication of a weather change, and I suppose I will just have to live with the pain the remainder of my life.

One thing that I can say is that had it not been for my wife, Judi, who was there daily from about eight a.m. to eight p.m. and my many family members and friends visiting and praying for me, I would not have made it through the ordeal. I had some friends tell me that I was too mean to die. I told them that was not why I survived. I said there were so many people praying for me that the Lord got tired of listening to them and just threw me back.

I was not sure that I would ever walk again, due to the damage of the accident in addition to the preexisting polymyocitis, for which there is no cure. I knew that I had become extremely weak over the three and a half months I had been in bed. I was so weak that I could not pick up a full quart water bottle.

On August 1, two therapists came to my room to get me up for the first time in three months. They brought a wheelchair into the room, wheeled it up to my bed, and said that I should sit up on the side of the bed. I tried to swing my legs off of the bed but found that I didn't have enough strength to do that. One of the therapists swung my legs off of the bed while the other helped me sit up. After sitting there for a few minutes, they said that they wanted me to stand up, turn ninety degrees, and sit down in the wheelchair. Try as I might, I could not put enough downward pressure on my previously broken arms to rise to a standing position. I didn't have enough strength in my arms or legs to do anything.

Realizing my dilemma, the therapists said that they would help me get into a standing position and into the wheelchair. They started raising me up off the bed, but the muscles in my lower back just locked up and would not allow me to straighten up to a standing position. After several tries, I lay back onto the bed, and the therapists said that we would try again the next day.

The next morning I was given a large dose of muscle relaxants with my breakfast, and the two therapists appeared again to get me up and into the wheelchair. This time they were successful in getting me into a standing position; however, when I tried to rotate ninety degrees to sit down into the wheelchair, I could not move my feet. Finally, with the assistance of the therapists, I rotated and sat down.

After they got my feet positioned on the footrests, one of them wheeled me around and out the door into the hallway. I had hardly gotten into the hallway when I was so overcome with emotion that all I could do was cry. I cried all the way down the hallway to the therapy room. A lot of the other patients just stared at me, but I didn't care because I was so happy to be off my back and out of that bed.

That first day in therapy, they first exercised my arms and legs. They then rolled me over to a horizontal bar, locked the wheelchair, and told me to reach out, grip the bar, and pull up into a standing position. I tried with all my strength but could not pull up. The therapists then got on both sides of me and helped me pull myself into the standing position. They told me to stand there as long as I could. Almost immediately I started to black out. They got me back into the wheelchair, moved me over to a lift table, and got me up onto the table. They strapped me down and started cranking the table in a vertical position. When I was just about vertical, I could feel the rush of blood to my legs and started blacking out again. They quickly lowered me back into a horizontal position and let me lay there a few minutes before cranking me up again, but more slowly this time. That time I didn't black out at all. After what seemed like an eternity, which actually was no more than two minutes, I was cranked down and moved back into my wheelchair, totally exhausted.

They moved me back to the horizontal bar and helped me stand, holding onto the bar. After a few minutes, my legs began to shake, and I was put back into my wheelchair. I was wheeled down the hallway, ready to get back into my bed. I had been out of my bed for a total of about thirty minutes.

The next day I went through the same routine of getting into the wheelchair and being wheeled down the hallway to the therapy room. This time I didn't cry from happiness, but I did have my life threatened by a female patient who was standing in the hallway. As I approached her, she threateningly glared at me and as I got closer, she said, "You don't think I can get you, but I can!" As I went by, she started moving her walker toward me. Thankfully, I had two therapists with me, who quickly moved me down the hallway.

When I got to the therapy room, I was put through the same routine as the previous day. I was able to do a little better and, at the conclusion of the exercise, thought that I was going back to my room. To my surprise, Kelly, the smallest of the therapists, shocked me.

She sat a walker in front of me and told me to stand up. I still didn't have the strength to stand by myself but was helped up by the therapists. I had never tried to use a walker before and didn't know what to do. Kelly told me to take one step forward, which I did. She then told me to pick the walker up and move it forward just slightly. I took another step, with difficulty, because I really could not feel my right foot, and repeated the move of the walker. I completed five repetitions of the move-and-step operation before my legs began to shake like two wet noodles. I was just about to fall when Beverly, the other therapist, pushed the wheelchair up against the back of my legs, and I literally fell into it. When I got back to my room, I was put into bed and collapsed into a deep sleep, totally exhausted.

The third and fourth days, I was put through the same routine as previously, except when I got to the walker. This time Kelly made me walk from one side of the room to the other. Also on the fourth day, Kelly had me sit on the corner of a padded exercise table. She told me to move myself from that corner to the other corner, about ten feet away, using my arms to lift my butt off the table and shift over toward the other corner. Since both arms had been broken and not yet fully healed, every push down against the padded table was extremely painful, to say the least. Kelly said that I was not going back to my room until I got to the other corner of the table. At that point, I didn't like her very much, even if she was cute. I finally got to the other corner and was helped back into my wheelchair. I thought I was going to my room but was shocked when Kelly moved me to the side of the room opposite the door, and placed the walker in front of me. She said, "Now get up and walk to the door."

After the ordeal on the table, I didn't have enough strength left in my arms to get up by myself. I was helped into the standing position and started the walk toward the door. I had to stop a couple of times, but I made it to the door and was allowed to sit in my wheelchair. Then Kelly took me back to my room and rewarded me with a cup of orange sherbet. Each day thereafter, I repeated that routine with longer and longer walks with the aid of the walker. I made sure that I was rewarded each day with a cup of sherbet.

From May to mid-August, 2004, the only thing I wore was a hospital gown. After that, I wore loose fitting pajama pants for months, because I could not tolerate anything tight over the surgical areas.

On August 11, I was taken to the Orthopedics Center near Huntsville Hospital for a routine check-up and x-ray of my arms and pelvic areas. While there, Dr. David Griffin, my primary doctor, asked if I was at home yet, and I immediately told him I wasn't. He continued with more questions about how I was walking and if I could get to the bathroom on my own. I assured him I could, with a little help. Dr. Griffin told me it was time for me to go home. He signed my release papers, and I returned to the nursing home. Judi and I kept the administration staff on their toes for the next few hours. They got me my own wheelchair and measured my paralyzed right leg for an orthopedic fixture. At four p.m., I was released and went home, happy and exhausted. It was like heaven sleeping in my own bed in our new home that night.

Long-Term Recovery

Almost immediately, I began in-home therapy to regain strength. I spent my days in bed, in my wheelchair, or reclined in my new lift chair. I could use a walker for short distances around the house and was required to walk several times a day. I used dumbbells to strengthen my arms because the muscles had atrophied.

By early 2005, I graduated from my wheelchair to using a walker full-time. I would walk further and further until I could walk almost a half mile before becoming totally fatigued. I continued in that mode of mobility for the better part of a year. We started making some short trips to Memphis to visit our daughter and her family and to Destin, Florida, for a few days vacation. I still had to take pain pills for the longer trips. Even today, I can't sit straight up for long periods of time without my pelvic areas hurting. I can get some relief from a reclining seat, because it takes some pressure off of the pelvic bones. In addition, I use a cushion to make the seat softer.

By summer of 2005, I had again graduated, this time to two walking sticks, mostly for indoor use. By the end of 2006, I could walk with just one. I use the stick on my right side where my leg is paralyzed because I, more or less, just throw that foot out in front of me and hope that it is on something solid before I pick my other foot up to take a step. I use a wheeled walker for outside excursions or on sloping areas.

Judi told me that while I was in the hospital, under the influence of strong drugs, I said some strange things, which I don't remember. Dr. Griffin told her that I might say or do something out of the ordinary after going home because of the lingering effects of the strong drugs.

This is where it gets a little strange. I am sure that most of you have heard about people relating near-death, out-of-body experiences. About two years after I came home from the nursing home, I started having flashbacks of hanging upside down in a brightly lighted operating room. I could see several people moving around me, all with white clothes on.

After several of these flashbacks, I asked Judi if they had me upside down while I was in the hospital. She said, "Yes, Dr. Griffin said that they had to hang you upside down to operate on your pelvic areas." I was shocked, to say the least. The whole thing may have been because I did not have enough anesthetic during the operations and became partially conscious, enough to realize I was upside down. I suppose I will never know the real cause.

Some friends asked me if I was going to sue the young lady that caused the accident. I had a considerable amount of time to think about that while in the nursing home and while recovering at home. I remember hearing the lady that caused the accident exclaim just about the time the vehicles stopped moving.

"Oh, God," she said.

I could plainly hear her because all windows were gone from my truck. I don't know whether that was a call for God's intervention into the matter or just a comment of fear and frustration on her part. At first thought, I was very angry with her for what she had caused. After much thought, I finally decided that if it was a call for God's intervention, then that may explain why I survived the accident. Anyhow, what's done is done, and that can't be changed. I decided it was not a good choice to cause possible ruin and despair to another person.

End of Story

Well now! I have rambled around through events of my life and come to the conclusion that I have had a good one. I just wonder where all the years went so fast. All of a sudden, I am an older man with grown grandchildren, very little energy, and a propensity to fall asleep at the drop of a hat.

In hindsight, I had a wonderful, interesting working career in the aerospace world that I feel I could not have obtained in any other field. I have heard others say that they became bored in their selected careers, but I can truthfully say I never was bored in mine in any way.

I was part of a team that accomplished some of the most significant events in the history of the world. We engineered and built the largest rocket (Saturn V) ever built by mankind. Using that rocket, we put man on the moon for the first time. We put the first space station in orbit (Skylab) also using that rocket. Then, we put the largest space station in orbit using the first re-useable space ship (the space shuttle) ever built. In addition, we laid the foundation on which future space activities can be built.

I am still amazed at the knowledge and expertise that was assimilated within NASA, private industry, and academia to solve the problems associated with NASA's missions of going to the moon and exploring the universe. Never before in the history of mankind had so many knowledgeable people come together for common causes. When I think about it, I am humbled by the fact that I was able to be a part of it all.

After World War II, the United States struggled for about fifteen years with the direction the nation should go. The Cold War with Russia was the focus point of everybody's thoughts. No one really knew where the nation was headed. Then President John Kennedy was elected. The Russians launched the first orbiting satellite, Sputnik, and Kennedy reacted. He announced the NASA mission to go to the moon with a timetable that the nation could support, and the whole nation prospered as a result. The nation became the world superpower with the technology developed by the effort to go to the moon.

I firmly believe that the politicians in Washington, D.C., have made some bad decisions relative to the US space program. I believe that they prematurely stopped the Saturn V program before gleaning all that could be accomplished with that vehicle, all for the sake of politics.

I'm afraid that our politicians today have lost the foresight to take the nation in the direction required for it to survive as the world leader in many fields. I believe that Congress should fund the NASA programs for long periods instead of year to year, because of the long development and implementation time requirements.

My career with NASA and all the technological advancements it entailed, in addition to what I have learned about this universe, has changed my perspective about this planet called Earth and of life itself.

I never was the best at anything. Throughout my life I forgot birthdays and anniversaries just like everyone else, and I have my many faults, but I tried to learn as I went and not repeat mistakes that I made.

While lying flat on my back in the nursing home after my accident, I had a lot of time to just think of the ups and downs of life. I think I have learned a few things over the years that would be good to pass on.

- In this world, there are givers and takers. Givers are usually those that are the leaders and those that are not afraid of taking risks. Takers are usually followers with less confidence in themselves and afraid to take risks.

- When my grandfather was still living, he often told me, "Don't believe anything you read, anything you hear, and only half what you see, because things are hardly ever as they appear." I believe he was a very wise man. I have seen things, previously taught as fact when I was in school, changed because of knowledge discovered during the space age.

- Learn to look at scientific achievements with an open mind, because most knowledge about the earth and universe that we live in has been learned in the past one hundred years, and we have only begun to learn.

- Just remember that millions of people, including those considered educated leaders of the world and church, once believed that the earth was flat and that everything in the universe revolved around the earth. Throughout history, many so-called facts have turned out to be wrong.

- History has shown us that there have been many people that kept an open mind, often thinking "outside the box," and have changed the world as a result. Consider the accomplishments of Michelangelo, Archimedes, Columbus, Werner Von Braun, and the founding fathers of our nation, just to name a few.

- Learn to accept others faults and choose to accept and enjoy other's differences. That is the key to a healthy, lasting relationship. Just remember this—when you point your finger at someone else, you have four pointing back at you.

- Tell people you love them at every opportunity. It may be your last chance, for in the frame of time in this universe, we are here only a fleeting moment.
- Put God first in your life, then family and then friends, and keep them in that order. Just remember, prayer is our most powerful asset.
- Don't hold on to anger, guilt, frustrations, problems, and sin. Just lay it at the foot of the Cross and leave it there. Don't ever go back and pick it up. Let Jesus, who has already paid the price bear your burdens.
- Surround yourself with things you like family, music, plants, hobbies, keepsakes, etc., to keep your mind busy.
- Don't take yourself so seriously; hardly anyone else does.

May God bless!

Appendix

Abbreviations:

AEA- Alabama Education Association
ATM- Apollo Telescope Mount
BAL- Blood Alcohol Level
CCC - Civilian Conservation Corp.
CID- Criminal Investigation Division of the FBI
C.S.A.- Confederate States Of America
CSRS- Civil Service Retirement System
ET- External Tank
FERS- Federal Employee Retirement System
GPA- Grade Point Average
GS- Grade Structure
HEAO- High Energy Astronomical Observatory
ICU- Intensive Care Unit
LAGEOS- Laser Geodynamic Satellites
LOX- Liquid Oxygen
LRV- Lunar Rover Vehicles

mm- millimeter

MPH- Miles Per-Hour

MSA- Marshall Spray Ablator

MSFC- Marshall Space Flight Center

NASA- National Aeronautics and Space Administration

OTC- Officer Training Course

PEC- Productivity Enhancement Center

ROTC- Reserve Officer Training Corp.

RN-Registered Nurse

RSRM- Re-designed solid rocket motor

SOFI- Spray-On-Foam-Insulation

SRB- Solid rocket motor segments with forward skirt, nose cap, and aft skirt

SRM- solid rocket motor

SRM&QA- Safety, Reliability, Maintainability and Quality Assurance

TISP- Teachers In-Space Program

TVA- Tennessee Valley Authority

UAB- University of Alabama Birmingham

UAH- University of Alabama Huntsville